WRITING MUSICAL THEATER

ALLEN COHEN
AND
STEVEN L. ROSENHAUS

WRITING MUSICAL THEATER

© Allen Cohen and Steven L. Rosenhaus, 2006.

First published in 2006 by
PALGRAVE MACMILLAN™
175 Fifth Avenue, New York, N.Y. 10010 and
Houndmills, Basingstoke, Hampshire, England RG21 6XS
Companies and representatives throughout the world.

PALGRAVE MACMILLAN is the global academic imprint of the Palgrave Macmillan division of St. Martin's Press, LLC and of Palgrave Macmillan Ltd. Macmillan® is a registered trademark in the United States, United Kingdom and other countries. Palgrave is a registered trademark in the European Union and other countries.

ISBN 1–4039–6395–9 paperback

Library of Congress Cataloging-in-Publication Data

Cohen, Allen (Allen Laurence)
    Writing musical theater / Allen Cohen and Steven L. Rosenhaus.
        p. cm.
    Includes bibliographical references and index.
    ISBN 1–4039–6395–9 (alk. paper)
        1. Musical—Writing and publishing. 2. Musicals—Production and direction. I. Rosenhaus, Steven L. II. Title.

MT67.C6 2006
808′.066792—dc22   2005050896

A catalogue record for this book is available from the British Library.

Design by Newgen Imaging Systems (P) Ltd., Chennai, India.

First edition: February 2006

10 9 8 7 6 5 4 3 2 1

Printed in the United States of America.

Contents

Permissions Acknowledgments

.

ACKNOWLEDGMENTS

We thank the following people, who have taught us or given us the support and love that helped us write this book: Beverly Chaney, Harold and Diana Cohen, Lehman Engel, Stella Kluger, Leo Kraft, Tsuya Matsuki, George Perle, Lawrence and Anne Rosenhaus, Ruth Rosenhaus, Bruce Saylor, and Hugo Weisgall.

We also thank our musical theater composition students, past and present, whose enthusiasm for learning their craft prompted us to write this book in the first place.

INTRODUCTION

This book is not only for people who want to write musical theater. It is for anyone interested in musical theater, anyone who would like a better understanding of how musicals are put together and how they work—or don't work.

When we started out as aspiring musical theater writers, we found no textbook or reference work to help us learn our craft. We were each fortunate enough to learn essential principles and techniques from teachers, writing workshops, and experience—but this remained a sporadic process, as it has been for most writers. When we began to write our own shows, we often wished we could find a book that analyzed musical theater from the writer's point of view, a comprehensive guide that presented the rules of the craft in a systematic and thorough manner. More recently, when we started to teach musical theater writing at the college level, we were still unable to find a suitable book for our students. Of the few books published on the subject, most are poorly organized and incomplete, and none discusses the music in any depth. At last we decided to write a book ourselves, like the one we wished we could have found.

Student and novice writers will find a great deal of practical information in this book, but it should be of equal value to more experienced writers, providing them with a set of basic principles and standards, as well as tips and suggestions, all within a general reference work. Writing musicals is a craft, or rather a set of several crafts, each with its own principles and techniques, and these crafts can be taught. Writers who wish to create musicals of quality need two things: an understanding of good craftsmanship, and practice. This book explains the former and offers opportunities for the latter.

Part I is an analytical survey, in which we separate musical theater into its various component parts and examine each of them. We discuss both general principles and specific techniques, and illustrate our discussions with many examples from the field.

Part II is a guide, a "how-to" tutorial that leads you step by step through the initial stages of creating a musical, based on the principles elucidated in Part I. Using two musical projects that we have created for this purpose as models, we take you through the writing process: finding the initial idea, developing the characters, working out the details of the story, planning the score as a whole, and starting to write songs. Finally, we point the way toward the later but equally important processes of rereading and rewriting.

At the end of the book are four Appendices with additional information. Appendix A is a brief history of musical theater in America; Appendix B contains lists of reference books and other tools for musical theater writers; Appendix C covers practical issues such as adaptation rights, collaboration agreements, agents, and possibilities for production; and Appendix D lists the classic works in the field, with which anyone who loves or wants to write musicals should be familiar.

In order to discuss music, we need to show it, as we do in chapters 6 and 8, and to assume that you know some basic musical terms like pitch, interval, tempo, the names of chords, and so on. But even if you don't know these terms and can't read music, you will still be able to understand almost everything in the discussions. We also assume that you have at least some familiarity with most of the classic musicals listed in Appendix D.

We'd like to make clear what we mean by *musical theater* in this book. We distinguish modern, commercial *musical shows*, or *musicals*, from plays with music on the one hand, and opera on the other.

Many plays such as Shakespeare's include songs and dances, but they differ from musicals in several respects. In plays, most of the running time is taken up by speech, while in musicals the balance between spoken portions and musical numbers is approximately equal. (Of course, this distinction does not apply to recent "pop operas," in which everything or almost everything is sung.) Another difference is one of function. In plays such as Shakespeare's *A Midsummer-Night's Dream* or *The Tempest*, the songs are mostly *intermezzi* or diversions; most of them could be cut without affecting the telling of the story at all. Except for revues (shows without a central plot line), the songs in a musical are part of the narrative framework; most or all of them further the telling of the story and the revelation of character.

Opera is also musical theater, and many operas are structurally indistinguishable from musicals. In both *The Magic Flute* and *Carousel*, for instance, the story is told through both song and dialogue, with some dancing as well. Again, however, there is a difference in the balance of spoken and sung words. While in plays the balance is weighted toward the spoken word, and in musicals the balance is usually equal, in operas the balance is usually tipped more toward the music. In a musical most of the emotional high points are sung or danced, but in opera, *all* of these points are musicalized, and most of them are sung. There is also a difference of expectations. Opera audiences expect to hear a traditional, unamplified orchestra and singers with highly trained voices. Musical theater audiences have become accustomed to smaller, amplified orchestras with nonsymphonic instrumentation such as saxophones, drum sets, electric guitars and basses, and electronic keyboards. They accept that the singers' voices may be amplified, and that those amplified voices may be less than operatic or even untrained. The dividing line is still hard to draw; Stephen Sondheim, whose shows have appeared both on Broadway and in opera houses, has said that when *Sweeney Todd* is performed in a Broadway theater it is a musical, but when it is performed in an opera house it is an opera. Much of this book will apply to opera as well as musical theater—but our focus is on the latter.

All of the concepts and techniques discussed within this book are based on experience and analysis. Many of them come from experience gained during the creation of our own shows, those of colleagues and friends, or other new shows with which we have been involved professionally. Many come from the experience of renowned professionals, from whom we have had the good fortune to learn. Our analyses have also built upon the wisdom of these veterans, as well as upon our own teaching experience. There are exceptions to almost everything, and all of the principles we state are generalizations, as we will occasionally remind you. But each one has a great deal of experience behind it.

One final thought: Whatever the economic and artistic state of musical theater may be, there is still need for excellence and innovation in the field, now more than ever. We hope this book will help aspiring writers to become leaders and innovators in tomorrow's musical theater.

Part I

The Elements

CHAPTER 1

THEATER BASICS

This chapter explains the basic facts and terminology of subjects such as theater spaces, stage geography, and types of musicals—information that anyone interested in understanding musical theater needs to know, and that is referred to throughout the book. If you are unfamiliar with these basics, you should read this chapter. Otherwise, you can skip it and proceed directly to chapter 2.

Stages and Theaters

TYPES OF STAGES

There are three common types of stages. In large commercial theaters the most common type is the *proscenium*. The word comes from Latin and means "in front of the scene," which refers to the rectangular arch that frames the front of the stage. The stage lies behind an opening in one wall of the theater space, with entrances and exits usually made from the wings (the sides of the stage). This rectangular opening between the stage and the audience is often called the "fourth wall," because when the stage setting represents a room in "realistic" plays and productions, the opening is treated as one of the room's walls. Sometimes the stage is separated from the rest of the theater by a curtain that is lifted—or by two curtains that are pulled apart—to reveal the stage for each act or scene. The proscenium arch is the frame for all of the action, and it allows a wide variety of scenic elements and backdrops. (A *backdrop* or *backcloth* is the scenic element at the rear of the stage, a piece of canvas that is usually hung like a curtain and painted. When used purely as a screen for lighting and sky

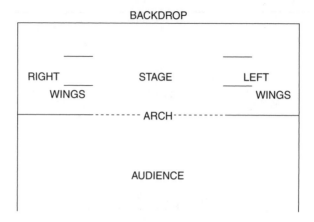

Figure 1.1

effects it is called a *cyclorama* or "cyke.") Figure 1.1 shows an overhead view of a proscenium stage.

The second most common type of stage is called *thrust*, and is possibly the most ancient of all stages; the classical Greek theaters had thrust stages, and Shakespeare's Globe Theatre also had a type of thrust stage. There are no traditional Broadway theaters with thrust stages, but several institutional New York theaters have them, as do many off-Broadway theaters. The ancient Greek theater was an open-air amphitheater with a circular stage; the audience sat in curved rows that extended about two-thirds of the way around it. In a modern thrust theater, the stage extends out from one wall (which is sometimes used for scenic backdrops), usually in a rectangular shape, and the audience sits on either side of it as well as in front (see figure 1.2). Entrances and exits are usually made from either side of the rear wall. There is no curtain and no frame, and usually there is little scenery except for some furniture, some set pieces in front of the backdrop, and the backdrop itself.

The third common type of stage is the *arena*, which may date back to the outdoor circuses of the Roman Empire, such as the Colosseum. There are no Broadway theaters with arena stages, and even off-Broadway they are rare, but they can be found off-off-Broadway, in many college theaters, and in many regional and outdoor summer "music circuses" around the country. The stage is usually in the center of the theater space, and entrances and exits can be made from any aisle or corner (see figure 1.3). There is no curtain and no frame, except the boundaries of the stage. While arena stages are square as often as they are circular, arena staging is often called "in the round." There is little or no scenery; even a piece of furniture will block sightlines from somewhere in the audience. But the stage seems close to the entire audience, and minimal

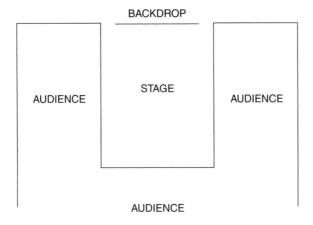

Figure 1.2

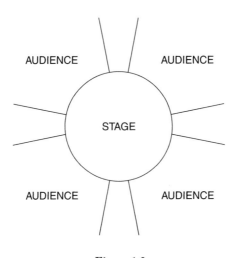

Figure 1.3

scenery leaves more to the imagination, which can make even small shows seem magical. The opening number of Tom Jones and Harvey Schmidt's *Philemon* pays tribute to the almost infinite possibilities available when drama is performed "Within This Empty Space."

There are other, less common types of stages and theaters that could be called "site-specific performance spaces." Some, like old nightclubs or dance halls, can be appropriate places for the presentation of musicals, as the old Studio

54 discothèque was for the 1990s Broadway revival of *Cabaret*. Others are clearly unsuitable for live music, such as when plays or performance pieces are staged throughout the rooms of a large building. This has never been done with a commercial production of a musical, because the problems of where to put the musicians and how to coordinate them with the singers would probably be insuperable.

THEATER SIZE

As defined in contracts with the theatrical unions, New York City's "Broadway" theaters have five hundred seats or more. This group is sometimes divided into two categories: large Broadway theaters with more than a thousand or so seats (like the St. James, the Winter Garden, and the Majestic), which tend to house big musicals with large casts, and smaller Broadway theaters (like the Golden or the Booth) which usually house straight plays or small-scale, intimate musicals.

Off-Broadway theaters have between 99 and 499 seats. New York also has a third, unofficial category of theaters called off-off-Broadway. These theaters have fewer than 99 seats, and are rarely used for commercial productions.

Experienced musical dramatists have at least some idea of the size of the theater for which they are writing. In most successful shows there is some correlation between the scale of the show and its theatrical home. In part this is a matter of practicality; for example, a big show with a large cast and a mammoth set like *Les Misérables* would be almost impossible to mount in an off-Broadway theater. On the other hand, *Godspell*, with its small cast and simple set, worked extremely well off-Broadway but closed abruptly after it transferred to a large Broadway theater. We use the term "scale" rather than size because it is not simply a practical matter of large casts or elaborate sets. Scale is difficult to put into words, but easy to feel in a theater. Many theatergoers have seen shows that seemed too small or intimate to "fill" the space of a large theater. The Broadway shows that get the harshest reviews from critics are usually not the worst-written shows, but the ones that feel too small for the theater in which they open.

The writers' conception of the show as a whole determines its scale, which is only tangentially related to the size of the theater. For example, Jones and Schmidt's *I Do! I Do!* had a cast of two, while the same writers' *The Fantasticks* had a cast of eight. But as Tom Jones himself has said, *I Do! I Do!*—with its stars Mary Martin and Robert Preston, its somewhat broad, emphatic style, and its fanciful dancing set—was a large-scale "Broadway" show, while *The Fantasticks*—with its unknown actors, its quiet poetic style, and its simple unit set—was a small-scale "off-Broadway" show. Of course these two categories are not necessarily exclusive. Using nonstar casts and simple sets, *I Do! I Do!* has played successfully in countless small theaters. But *The Fantasticks*, even with stars in the cast, has rarely played successfully in a large theater; it is simply too small and intimate a show.

Yet neither physical size nor the size of the budget has a direct correlation with emotional scale or power. There have been many big, elaborate Broadway productions that left audiences cold, while *The Fantasticks*, in a tiny theater in Greenwich Village, charmed audiences for decades. Indeed, it was more effective, more magical, in that tiny space than anywhere else.

STAGE GEOGRAPHY

When talking about an ordinary proscenium stage, theater people commonly refer to a nine-cell grid in which the stage floor is split in thirds from left to right and from front to back, giving each cell of the grid a name: up right, up center, up left, stage right center, center stage, stage left center, down right, down center, and down left (see figure 1.4). (In the theater, unless otherwise indicated, "right" and "left" mean "stage right" and "stage left"—that is, right and left as seen from the stage facing the audience; stage directions are always from the viewpoint of the actors, not the audience. "Up" means toward the rear of the stage, and "down" means toward the audience.) Sometimes numbers are used to denote the front-to-back split of the stage, including the offstage wings, with the number one denoting the downstage area, two the area behind it, and so on. This is the origin of the old term "in one" to denote scenes or numbers that were done in the frontmost area of the stage with a curtain shutting off the rest, usually to cover a scene change. Because scene changes have become faster and more fluid, the need for scenes or numbers "in one" has disappeared, but many famous songs began that way, such as "There's No Business Like Show Business" from *Annie Get Your Gun*.

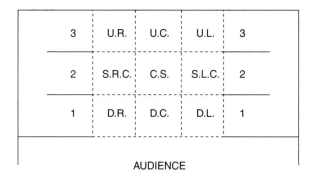

Figure 1.4

SCENERY, PROPS, LIGHTING, AND COSTUMES

Ordinarily writers don't need to know about the technical aspects of theater production; the people who specialize in them know far more than writers will ever need to know. A playscript usually includes no more specific scenic or lighting directions than are absolutely essential to set the scenes, no matter what the style of the show may be.

Once in a while there may be a place in a script that calls for a specific scenic effect. But usually the script will only describe the effect desired, not the means to achieve it. The means will inevitably change as the technology changes. Most other technical terms are similarly unnecessary for the writer to know.

The most common references to lighting in a script are cues to indicate the end of a scene. The most common one is "lights down" or "lights out," in which the speed of the dimming of the lights is assumed to be moderately fast. Sometimes a specific effect is called for: "lights out quickly" or "blackout" for a fast ending, or "the lights fade slowly" for the opposite. When the end of a scene has no lighting indication, it usually implies either a "lights down," or a crossfade to the next scene. Each director will have his or her own ideas.

Experienced writers usually try to avoid what Lehman Engel, the celebrated Broadway conductor and teacher, called "green sequin numbers." A green sequin number is one that needs specific sets, lighting effects, props, or costumes in order for it to work, so that when the songwriters present it to anyone they must preface it with a detailed explanation: "You have to understand that when she sings this number, she's wearing a fabulous dress with green sequins and an eyepatch, and standing in a purple spotlight under a crescent moon. . . ." A green sequin number, even a good one, is a potential liability, because it is hard to present outside the show—for instance, when trying to interest a producer. Sometimes it is hard to present *in* the show, when the budget is smaller than the writers anticipated, and nowadays many shows go through smaller productions before they hit the "big time." Because of the required effects, a green sequin number is also less likely to have a life after the show in recordings, revues, club acts, and so on.

ORCHESTRAS

As with scenery, props, and lighting, the details of the arrangement of the orchestra and the orchestra pit are not usually needed for the writing of musical theater. There are certain exceptions, of course, one being the need for everyone in the orchestra to be able to play their instruments and see the conductor.

Another is the necessity for the conductor to see the singers, and vice versa, for cues and tempos. And sometimes the conductor needs to be visible to the audience, either for comedic purposes (as in certain parts of *Chicago*) or for dramatic ones (as in the chaos sequence at the end of the original version of *Follies*).

In many recent shows the orchestra has been backstage, or offstage right or left; sometimes it is in the orchestra pit but the pit is covered, as in *A Chorus Line*. In these cases, the only visual contact between the conductor and the singers is through television: The conductor has a monitor in front of him showing the stage, and three or four monitors showing the conductor are hung on the front rail of the balcony so that the singers can see his image when they look out front. Any changes in the position of the orchestra add to the problems of transmitting the sound of the orchestra to the audience, and balancing it with the sound of the singers. Often these changes are not due to the script but to directorial concepts or practical constraints.

Thrust and arena stages pose even more problems for musicals. The biggest problem in such theaters is where to put the orchestra. Sometimes in thrust staging the orchestra is under the stage; often it is on a platform above or at the back of the stage, which poses additional problems of acoustics and coordination. Arena theaters often have an orchestra pit running along, or under, part of the circumference of the stage, but again this makes it inevitable that from many angles the singers will not be able to see the conductor—and there is usually no place to put television monitors for them. Ways have been found around these problems, especially with modern video and audio technology—but the bigger the show and the orchestra, the bigger the problems. For instance, when the musical *Starlight Express* came to Broadway, because of its basic concept—performers on roller skates skating around the theater pretending to be trains—it was decided to put the entire orchestra in a room on an upper floor of the theater building. Thus the conductor had to watch the action on a video monitor, and all the sound had to be transmitted between the theater and the orchestra room. Several times the curtain had to be held because there was a problem with the cables between the room and the theater. Until it was fixed, the orchestra could not be heard in the theater!

Decisions about the size and constitution of the orchestra can almost never be made before an actual production takes place. Even then, while a fortunate composer may be given a say in the question of orchestra size, it is really the province of the producer, and it is often dictated by union rules. As for which instruments will be included in the orchestra, this depends to some extent on the writers' concept of the show, but often it is dependent on the director's concept as well. In any case, the task of arranging and orchestrating the show is almost never done by the composer, but by another musician hired specifically for that task. This is not only because many composers don't have these skills, but because of the pressures of time. The best key for each song (except chorus numbers) can only be chosen definitively after the singer has learned and rehearsed it, and—given the realities of rehearsal time—there is at most six weeks for the entire score to be arranged and scored. The composer is usually much too busy with rehearsals and rewrites to do this mammoth job as well. The only Broadway composer who did all of his own orchestrations was Kurt Weill.

Types of Musicals

BOOK SHOWS

A "book musical" has a central story line and a consistent cast of characters. (The text of this type of musical is its *book* or *libretto*.) Traditionally the story is told as a linear narrative in which speech, song, and dance alternate smoothly and unobtrusively. The musical numbers usually occur at crucial points in the story—that is, at emotional high points. They are also often found at significant structural points, such as the beginning and the end of each act.

The book show has always been the most common type of musical. It includes every one of the most popular and acclaimed shows of Broadway's "golden age," a period that began in 1943 and lasted for about twenty years. Within this category there are three special cases worth mentioning.

The first special case is known as the "concept musical." Concept musicals differ from other book musicals in that, as a rule, speech and song do not alternate smoothly. Instead, the differences and the seams between speech and song are emphasized, to deliberately create a disjunctive effect. In a concept musical the songs stand outside the spoken scenes; they comment in some way, often with irony, upon the story that they interrupt. The show as a whole is usually arranged according to some overall concept or metaphor. For instance, because both of the lead characters in *Chicago* are performers whose dream is to headline in vaudeville, the show tells its tawdry story of infidelity, greed, and murder in the "Roaring Twenties" as a series of vaudeville acts. Sometimes the narrative of a concept musical is not linear but fragmented or recursive, as in *Company*. Because of their discontinuous nature, concept shows are usually less emotionally involving, and less successful with audiences, than traditional book musicals.

Love Life, from 1948, is often considered the earliest concept musical. (Some people would nominate *Lady in the Dark* from 1941 or *Allegro* from 1947.) The German musicals by Bertolt Brecht with Kurt Weill and other composers, such as *The Threepenny Opera*, are often cited as immediate ancestors.

But the genre truly flourished in the late 1960s and 1970s with such shows as *Cabaret, Company, Follies, Chicago,* and to a lesser extent *A Little Night Music* and *Pacific Overtures. Cabaret,* the earliest of these, could be considered a tentative step from the traditional book musical toward the concept musical: The score is split almost evenly between the concept numbers in the Kit Kat Club and the more traditional songs within the scenes, such as "So What?" and "Meeskite." When Bob Fosse directed the movie of *Cabaret* several years later, he eliminated the songs within the scenes, leaving only the Kit Kat numbers. Unlike the original show, the movie is a pure concept musical.

The second special case is the "through-composed" musical. The term comes from art music; when applied to an opera, it means that the music continuously flows and develops without any break from the beginning to the end of each act—as in the late dramas of Richard Wagner—rather than being constructed from discrete songs with repeated sections or stanzas. Commercial musical theater is never truly through-composed, but the term is often used loosely to denote shows with little or no dialogue, such as those with scores by Andrew Lloyd Webber or his imitators. Through-composed musicals differ from other book shows in that they are entirely or almost entirely sung, and even the brief spoken portions are usually accompanied by musical underscoring. *The Golden Apple* of 1954 is an early isolated example, but more direct ancestors are the recorded "rock operas" of the late 1960s such as The Who's mini-opera "A Quick One" and their full-length *Tommy.* The genre really came into its own in the 1970s beginning with *Jesus Christ Superstar*—which, like *Tommy,* was an audio recording before it had a stage production—and such shows are still sometimes called rock operas, "pop operas," or "poperettas."

The third special case can be called the "anthology" musical, which is simply a group of short book musicals that are intended to be performed in a single evening. The component shows are usually linked by a common theme. Anthology musicals are rare; the best-known examples are *The Apple Tree* and *Romance Romance.*

REVUES

The *revue* is a show without a narrative story line or a consistent roster of characters. It can include scenes as well as songs, and the songs can be either within scenes or self-contained. The revue is at least as old as the book show, with antecedents stretching back through vaudeville and variety shows to minstrel shows and beyond. There have been many types of revues in the history of musical theater. The three most common types were large-scale extravaganzas like the Ziegfeld *Follies;* variety shows built to feature particular stars such as

Bert Lahr or Beatrice Lillie; and shows built around a particular theme or concept, like *As Thousands Cheer, Pins and Needles*, or *This Is the Army*. All three of these types have virtually vanished, although *Bring in 'Da Noise, Bring in 'Da Funk* is a recent revival of the thematic revue. In the last few decades, almost all successful revues have been retrospective anthologies. Most of these featured a particular songwriter or songwriting team, and two of them, *Jerome Robbins' Broadway* and *Fosse*, surveyed the work of a director-choreographer.

MUSICALS FOR CHILDREN

The musical for children is often put into a category of its own. It is usually a book musical in one act, and no more than one hour long. (This category does not include Broadway book shows with children's themes such as *Peter Pan*, *Annie*, or *Beauty and the Beast*, which we will discuss in chapter 3.) There are at least two general types. Shows for very young children tend to be based on fairy tales or fantasies. They use simple language and songs, a good deal of action and broad physical comedy, and often some degree of audience interaction, such as a character asking the audience for help in finding another character. Shows for older children, on the other hand, tend to be much more verbal, and to have as the subject matter either the life of a famous person or a real-life situation common to the target age group. Often these latter shows are extremely similar in storytelling technique and craftsmanship to regular book musicals. The main differences lie in the subject matter, the sophistication and frankness of the language, and the length.

NEW GENRES

In recent years, there have been a number of musicals that do not fit into any of the traditional categories, but may be the beginnings of new ones. One new genre could be called the "cover" or "jukebox" musical, in which old songs appear within a new story. Examples include *Mamma Mia!*, which uses songs recorded by the pop group ABBA to tell an original story, and *A Class Act*, which uses theater songs written by Edward Kleban to tell the story of his life. (*George M!* from 1968 is an early show of this type, with songs by George M. Cohan.)

Another new genre might be called the "dance" musical—not a revue, but rather a modern equivalent of the classical ballet, a show that tells a story (or stories) through dance. An example is *Contact*. Because it has no singing, it is not truly a musical in the usual sense of the word, but it was marketed as one.

Another example of this genre is *Movin' Out*, which tells a story by combining pop songs by Billy Joel, sung as narration, with choreography by Twyla Tharp. While these shows were reasonably successful, neither the dance musical nor the cover show is very relevant to songwriters, because both types are almost always based on pre-existing songs.

There are other recent "movement shows" that have little or no dancing, but rather choreographed movement with or without musical accompaniment. The best-known of this genre is probably *Stomp*, although new variants continue to appear. With these hybrid genres, however, we approach the "no man's land" between musical theater and other types of "performance art." It's safe to assume that as time goes on, other genres will evolve and emerge.

Because children's musicals are essentially book musicals, revues other than retrospective anthologies are rare, and new genres by their nature are hard to define or discuss, *Writing Musical Theater* will focus on book musicals. The book show is by far the most common genre of musical, and many of the principles of dramatic structure that we discuss will be applicable in obvious ways to the other genres.

ADAPTATIONS AND ORIGINALS

Book shows can be classified as either *adaptations*, which are based upon dramatic or literary works, or as *originals*. An adaptation often provides the writer with a ready-made, usable plot and characters. In addition, the plot and characters of an adaptation have already demonstrated their appeal. An original requires more effort to cast in dramatic form, because it has not already been a dramatic or literary work. But originals have two tremendous advantages over adaptations. First, there is ordinarily no necessity to investigate, or pay for, the rights to the material. Second, a musical based on an original idea is the first presentation of the idea or story in dramatic form, and does not have to compete with a successful original version.

Chapter 2

The Idea

An idea is the genesis of every work of art. Often the idea for a musical comes from a movie or play. Sometimes the inspiration is a book, as Washington Irving's *History of New York* was for *Knickerbocker Holiday*; sometimes it is a painting, as Georges Seurat's *Sunday Afternoon on the Isle of La Grande Jatte* was for *Sunday in the Park with George*; and sometimes it is a poem, as Homer's *Iliad* and *Odyssey* were for *The Golden Apple*. The idea for a musical may also come from incidents in a real person's life, as Eva Peron's career inspired *Evita*; from historical events, as the history of Japan's foreign relations from 1853 to 1976 inspired *Pacific Overtures*; or from a satirical perspective on current events, as the Elvis Presley phenomenon inspired *Bye Bye Birdie*. The idea for a revue can come from the desire to showcase the work of a particular songwriter or choreographer, or simply to entertain an audience.

Why a Musical?

Before we take a look at the kinds of ideas that make good or bad musical theater, let's consider a more fundamental question: Why write musical theater at all? In other words, what is special about this hybrid of drama, song, and dance? Does it offer an aesthetic experience that cannot be found in as potent a form in any other genre? And if so, what kind of experience is it?

Musical drama seems to be as old as drama itself. We know that the plays of the ancient Greeks were not simply spoken, but were sung and danced as well. Chinese drama is at least eight hundred years old, and only in the last century has it included spoken plays. Throughout the Middle Ages, traveling performers presented combinations of theater and music. Opera was invented in Florence shortly before 1600 by a group of writers who deliberately modeled their works on the Greek tragedies. And modern musical theater, which is usually given a birth date of 1866, is still going strong after one hundred and fifty years.

Clearly, musical theater of some sort has been one of the most popular art forms in history. As with any popular art form, however, different aspects of musicals appeal to different people. Some people love musicals for their fantastic aspects, the spectacle and the glamor. Some like striking dance routines with pretty girls or boys. Some like to hear beautiful or powerful singing.

But while all these things are prominently featured in many musicals, and none of them can hurt, none of them is essential. For example, there is neither glitter nor glamor in shows like *West Side Story, Fiddler on the Roof,* and *Little Shop of Horrors.* There is virtually no dancing in *You're a Good Man, Charlie Brown* or *A Little Night Music.* And as important as singing has always been to musical theater, there are only a few opportunities for great singing in either *The King and I* or *A Chorus Line.* So when we talk about what's essential to musicals—what makes the good ones good—we are referring to something else that these and all the great musicals have in common.

Our fundamental premise is that the essence of musical theater is the representation of human emotions onstage, and the evocation of emotions in the audience, through the union of drama and music. This union may include spectacle, and often includes dance or choreographed movement; but the unique aspect of musical theater, whether dramatic or comedic, is the heightening of the emotional impact of a story or idea through music and song. We believe that there can be no great musical theater without strong emotion. What is the point of adding music to drama unless it is to embody, enhance, and illuminate emotion? Professionals in the arts know that music is one of the most powerful ways to enhance the emotional content of a story, which is why almost every film ever made, silent or talking, has had musical accompaniment. For most of us the great operas and musicals are the ones in which we have not learned things, but *felt* things—shared other people's pain or joy, laughed at other people's foibles—as confirmations or illuminations of our own feelings.

Many people love the excitement and razzle-dazzle, the pizzazz of big Broadway production numbers with dozens of dancers and flashing lights. But razzle-dazzle is a momentary phenomenon, and without something emotionally meaningful to sing and dance about, it's just glitz and sequins, "only a paper moon."

Our next premise is that writers who cherish musical theater at its best seek to emulate its best writers. The great theater songwriters and librettists—such as Jerome Kern, Irving Berlin, Cole Porter, the Gershwins, Lorenz Hart,

Rodgers and Hammerstein, Frank Loesser, Lerner and Loewe, Leonard Bernstein, Bock and Harnick, Stephen Sondheim, Arthur Laurents, and Terrence McNally—have aimed high, holding themselves to a standard of originality in conception and excellence in craft, and resisting the temptation to get away with slipshod workmanship. Even if some people *have* gotten away with it, there is no reason for new writers not to try to be their best. Nor would there be any point to a book that didn't hold up the best as the standard against which writers need to measure themselves. In this book we focus on, and use as models, examples of musical theater at its best.

What Makes a Good Story for a Musical?

The resumé of every experienced theater professional includes flops as well as hits. If even the most talented and experienced writers have had unsuccessful shows, then clearly talent and experience are not enough to guarantee a success, either artistic or commercial. There is a different story behind every unsuccessful show, but bad shows usually happen for one of three reasons.

The first reason is incompetent execution. This type of failure is rare, now more than ever. There are more professionals singing, acting, dancing, directing, playing, and designing musical theater than ever, and there are fewer opportunities. So the level of performing and designing talent available to a first-class production is extremely high. It is true that shows have been ruined by directors who were incompetent or simply unsuited for the material, but even this is comparatively uncommon.

The second reason is bad writing, or insufficient talent in the writing. This is a much larger category. There are many talented people who are writing musicals and trying to get them produced, but not everyone who gets produced is talented. As in every other artistic field, bad musicals tend to outnumber the good ones by far.

The third and most heartbreaking reason for failures is that good writers make bad choices. Since experienced writers have usually learned to avoid mistakes in the techniques of stagecraft and songwriting, we venture to say that the most common bad choice made is the most important and basic one: the choice of a project.

A desire to write musical theater and an idea are not enough. The question remains whether the idea should be made into a musical—whether it can even work as a musical. If the writers do not have strong reasons to believe that the idea is better as a musical than as a play, a novel, a film, or something else—that having it performed onstage with singing and dancing will enhance the idea rather than dilute or ruin it—then they would be wise to forgo it and look for something else.

But what makes a good subject for a musical? Although there are no absolutely right or wrong answers, there are some subjects that are inherently more suitable for musical theater than others.

Since musical theater is about the expression and enhancement of emotion, it follows that stories that contain and evoke strong emotion, serious or humorous, are more suitable for musicalization than those that do not. As for emotional content, there are two general requirements. First, the emotions must be strong enough that it feels appropriate for the characters to sing. In the words of producer Stuart Ostrow, the most important question the writer must answer is whether the story "sings": "Will a song add a deeper understanding of character or situation?"

The second requirement is that the story must contain enough emotional content for an audience to care about the characters and be willing to follow them to the end. As Oscar Hammerstein II advised the young Stephen Sondheim: "I want you to say: 'Can I interest an audience in this to the extent that I am interested in it?'" Writers must consider what it is in the idea that appeals to them, and how likely it is to appeal to large numbers of other people. This has nothing to do with a story's milieu or subject matter. Before shows such as Oklahoma!, West Side Story, or A Chorus Line came to Broadway, few people thought that audiences would come to see a musical about cowboys and their girlfriends, or urban youth gangs, or the childhoods of show dancers. It is not the subject matter but the emotional content and skillfulness of its treatment that determine how good a show will be.

Some say that, because of the larger scale that music adds to emotional situations, the characters and the situation must be "larger than life." But the characters in Oklahoma! are not particularly larger than life, nor are those in Carousel, Brigadoon, West Side Story, A Chorus Line, or many other shows. Rather, they are amenable to larger-than-life treatment. The characters are presented as distinctively individual, yet universal enough for many people to identify with them. Each major character is depicted imaginatively and vividly. As Hammerstein said, the smallest story can feel important "if the characters are examined closely enough . . . and if the narrative of the incident is told with enough depth and human observation."

It follows, then, that a bad idea for a musical is one in which there is not enough emotion to sing about, or in which the audience does not care enough about the characters. Let's consider the potential of a classic play, The Front Page by Ben Hecht and Charles MacArthur, for musicalization. It is tremendously funny, and it has an interesting, fast-paced plot with many serious moments as well. There is also a certain amount of pathos and emotion in the situations of two minor characters, Molly Malloy and Earl Williams. But consider the two main characters: Hildy, a reporter, wants to get married and leave the newspaper business for good; Walter, his boss, wants him to stay because he is Walter's best reporter. They are friends, of a sort, but that is the extent of their relationship and their conflict. The play is a combination of screwball comedy and a satirical portrait of the newspaper business. There is little at stake for the main characters, and very little to sing about. (There has been a musical version of *The Front Page* which, unsurprisingly, was not successful.)

Another criterion that separates good ideas for a musical from bad ones is timeliness. No matter in what period its story is set, a commercial musical must be of its time. This timely quality will be reflected in many aspects of the writing, including the language and the musical style, but above all in the choice of subject matter and the way it is treated. In addition, there is a certain indefinable flavor of the *Zeitgeist* that a show must have in order to be right for the moment and to connect with a large audience. In the 1940s, for example, many people responded to tender romances in which love transcended time and space, such as *Carousel* and *Brigadoon*. But if the same shows were new today they would not last a night, despite their quality. Nor would a new show that treats as comedic material a man who wins back his ex-wife by tricking, humiliating, threatening, and striking her; but in 1948, audiences loved the same situation in *Kiss Me, Kate*. On the other hand, a show that is too dependent upon current events, ideas, and buzzwords, will be out of date by the time it closes, and possibly by the time it opens. If *Carousel, Brigadoon*, and *Kiss Me, Kate* still live sixty years later despite their dated qualities, it is because they deal with concerns like love, hate, greed, pomposity, and frustration—human feelings that are always with us.

There is another crucial criterion that applies to adaptations. Because an adaptation has to compete with the original on which it is based, a successful adaptation must enhance the story in some way not present in the original version. This can only be the case if the original is not already in its most suitable medium, and if a new version with songs and dances can add to it something better, or at least different. The best musical adaptations have rarely been based on the best plays or movies. The writers of *Oklahoma!* took the play *Green Grow the Lilacs*, made many changes to the story and the characters, and added a ground-breaking approach and a first-class score. In the process, a pleasant but undistinguished play was turned into a startlingly novel, record-breaking musical. With a first-rate play, there is less for an adaptation to improve or to add.

Let's consider two apparent exceptions. *My Fair Lady* is based on George Bernard Shaw's play *Pygmalion*, and indeed a small number of people have always felt it to be inferior to the original. But the musical adds a crucial element to the play that justifies the adaptation for most people: The new ending holds out the possibility that things will work out between Higgins and Eliza, and thus introduces the element of romance, albeit a cautious and understated one. Romance implies emotion, more emotion than is evident in Shaw's witty but dry original. It is impossible to imagine the Eliza of *Pygmalion* saying "I Could Have Danced All Night," but for the Eliza of *My Fair Lady* it marks a necessary step in her journey. Similarly, *West Side Story* would seem to invite an invidious comparison with its original source, *Romeo and Juliet*. But while the basic story was retained, almost everything else was changed. In addition, by eliminating all of Shakespeare's original language, the adapters avoided the trap of having their dialogue and lyrics compared to Shakespeare. A musical adaptation does not necessarily have to be *better* than the original, but it needs to be at least as good, and *different*. It must add something new to the original concept besides song and dance.

An example of a show that did not accomplish this, and as a result was only marginally successful, is *Raisin* from 1974. It was based on Lorraine Hansberry's *A Raisin in the Sun*, one of the most successful and powerful American plays of the mid-twentieth century. In 1974 the original play was somewhat dated but still relevant, with well-drawn and interesting characters and a compelling story full of strong emotion; the libretto of the musical was extremely faithful to the original play; and the score was high in quality and well suited to the milieu of the show. It must have seemed like a perfect idea for a musical. The problem was that the adaptation neither added nor changed anything fundamental. As good as the songs were, most of them were simply slotted into the story by replacing the equivalent original dialogue, with nothing significant changed. The libretto was very faithful to the original play, but it would not be far off the mark to say that *Raisin*'s basic problem was that it was *too* faithful. (Lehman Engel once wrote that one of the worst things you could say about an adaptation is that it is "faithful to the original.") But the crucial mistake was probably the decision to adapt a play as well written as *A Raisin in the Sun* in the first place. More recently, a similar problem afflicted *The Full Monty*, which, despite a skillful adaptation and an upbeat score, added nothing new to a delightful movie except a change of locale from Britain to America; this resulted in a pleasant but undistinguished musical. It would be even more misguided to musicalize such great plays as *Death of a Salesman*, *A Streetcar Named Desire*, or *Long Day's Journey into Night*. To make significant changes to the plays would be foolish because they are so good, but without changing them there would no point to adapting them at all. Professionals sometimes use the term "why musicals" to describe shows that either use a property from another medium without adding anything new, or have subjects unsuitable for musical adaptation.

Of course, what seems a good idea to one writer may seem an awful idea to the next, and both may have legitimate reasons. Before Lerner and Loewe turned *Pygmalion* into *My Fair Lady*, the chance to adapt Shaw's play was turned down by every major theater songwriter and team, including Rodgers and Hammerstein, Noel Coward, Cole Porter, Schwartz and Dietz, and even Lerner and Loewe! Some felt that it was already "a perfect work of art," and others felt that it posed insuperable problems for adaptation. Similarly, despite the reputations of Jerome Robbins, Leonard Bernstein, and Arthur Laurents, many producers turned down *West Side Story* before it came to Broadway. Some felt it was too much like an opera; some found the subject matter distasteful; and some felt that it was simply too depressing.

Still, there is a clear difference between shows like these, which "pushed the envelope" by violating the conventions of the time, and shows that lack the essential elements of a good musical in any era. For example, in 1943 it was considered necessary for every musical to open with a big splashy chorus number, and impossible to have a major character killed on stage—two conventions that Rodgers and Hammerstein's *Oklahoma!* rejected. (After seeing *Oklahoma!* in out-of-town previews, producer Mike Todd supposedly sent a telegram to a friend in New York that read, "No gags, no girls, no chance.") The big opening chorus number turned out not to be truly essential. *Oklahoma!* broke many precedents, but by its success it established new ones of its own, such as the parallel subplot and the dream ballet—that in turn were broken by later shows.

The true essentials of a good show, such as the need for sympathetic characters and a story with strong emotion, have not been superseded in the 150-year history of musical theater. In fact, they have changed very little in the 2,500-year history of musical drama. Stephen Sondheim quotes playwright and wit Wilson Mizner, who said "People beat scenery." Sondheim adds, "That's what the musical theater is about."

CHAPTER 3

THE LIBRETTO

The libretto of a show consists of all the words. *Libretto* is a term from opera, an Italian word meaning "little book," and the libretto of a musical is often called the *book*. (The Italian plural is *libretti*, but in English they are usually called librettos.) It is important to emphasize that the libretto is much more than just the dialogue; it also includes all of the stage directions and the descriptions of the sets. More important is that the libretto includes the *structure* of the show as a whole. It also specifies the relations between the scenes and the songs: what is spoken, what is sung, and what is danced, as well as the order of scenes. In the words of Peter Stone, author of the librettos of *1776* and many other shows, "a musical book is, in one word, *construction*."

As we shall see, librettists, or "bookwriters," often make significant contributions to the score. Songwriters contribute to the libretto just as often. Songwriters have to set the story to music, and they need to live and breathe it just as much as the librettist does. Since they make an essential contribution to the dramatization of the material, they necessarily take part in the writing of the libretto. Knowledgeable songwriters don't think of the libretto as the stretches of dialogue between the songs; they think of it as the totality of the show, *including* the songs.

As this chapter will make clear, the libretto is crucial to the success of the show, commercially as well as artistically. In the old days, hit songs could generate enough interest to bring in an audience despite bad reviews, but since the late 1960s it has been extremely rare for theater songs to get radio or television airplay. A great score can't keep a show going—but a great libretto can.

Dramatic Essentials

WHAT MAKES DRAMA?

The question of the essential nature of drama has filled many books and symposia, but it is safe to say that drama requires the representation of a strong desire and a conflict. According to an old adage, all that is necessary for drama is "four boards [to make a stage] and a passion."

Although language is an important aspect of almost all drama, it is not a fundamental necessity. In fact, it is possible for all or part of a dramatic work to have no language at all, as in ballet or pantomime. As the *Encyclopedia Britannica* has put it, "Drama begins with action and spectacle; the deed comes before the word, the dance before the dialogue, the play of body before the play of mind." More than language, drama requires *action*—which can be psychological as well as physical—and action almost always implies *conflict*.

Classical writers believed that there are three basic types of conflict. The first is the conflict of man against nature (or the gods, or Fate). In *Oedipus the King*, the central conflict could be said to be between Oedipus and destiny or Fate, while in *Carousel* it could be said to be between Billy Bigelow and mortality. The second conflict is that of man against man. In *King Lear* the central conflict is Lear against his daughters, or more generally, the old against the young, while in *1776* it is John Adams against his fellow Continental Congressmen. The third basic conflict is that of man against himself. Thus in *Hamlet* the central conflict is Hamlet against himself, while in *Jekyll and Hyde* it would be Dr. Jekyll against Mr. Hyde (who is the worst part of Dr. Jekyll).

All three types of conflict can make good theater, but generally the most powerful or touching moments in drama (and often the funniest as well) come from the third type, a single human soul in conflict with itself. If you doubt this, ask yourself which is more emotionally powerful: Othello in Act One, defending himself against the charge of witchcraft brought by his father-in-law—or Othello in Act Five, about to kill his wife, wavering between his jealous fury and his love for her? Is it Tevye in the next-to-last scene of *Fiddler on the Roof*, telling off the Russian constable—or Tevye in the last scene, torn between the stern traditions of his people and his love for his daughter Chava, who has rejected those traditions?

Since drama is action, and since it is always explicitly or implicitly about human beings, we can say that action and conflict occur, and character is revealed, through human behavior. When a good playwright shows us behavior, it is always *specific*; it shows characters in conflict—each taking, or trying to take, a specific action in pursuit of a specific goal. "Following your dream" is vague, and gives the actor nothing to act. "Convincing a wealthy man to loan you the money you need to start your business, by showing him how it will benefit him" is more specific. In each scene, and in the play as a whole, characters must either achieve their goals or fail. If the characters and situation are specific and interesting, then the audience knows exactly what it wants to see happen, or not happen, on stage. Good actors usually analyze their parts to determine the objective of their character within each scene, and the overall objective in the play as a whole.

There are more compelling reasons for specificity beyond helping the actors prepare their roles. Characters must show individual, particular ways of talking and behaving in order to convince the audience that they are people with their own lives and backgrounds, habits and mannerisms, needs and desires—like real people. For characterization to be specific, the characters must have specific modes of behavior—specific physical mannerisms, modes of speech, turns of phrase, and so on. This is true even if the characters are not supposed to be human beings but Greek gods, man-eating alien plants, or Snoopy the dog. Classical tragedy usually presented archetypal characters like gods and heroes; they and their stories were well known to their audience. But modern drama usually deals with particular characters in particular situations, and this is even more true of modern musicals. In today's fragmented, multicultural, heterogeneous society, it is easier for an audience to identify and empathize with specific characters than with symbols or archetypes, few of which are as universally known and meaningful as the Greek gods and kings once were.

For most actors, drama is fundamentally about not what the characters *say*, but what they *want*. In real life, these two things are often the same. But it's usually more interesting dramatically if what a character says is different from, or unrelated to, what the character really wants. This contradiction between what a character says and what he really thinks or wants is *subtext*. For an example, look at Curly and Laurey's first scene in *Oklahoma!* They are both too proud or too contrary to admit their feelings for each other. They insult and banter with each other, but that doesn't reflect how they really feel—and however contrived this banter now seems, it parallels many real-life situations in which people conceal what they are thinking or feeling. In short, there is a disparity or conflict between their thoughts and their actions or words, and that conflict makes for interesting human behavior.

Every drama has a shape, like any art that unfolds in time. Here we do not mean the division of a play into acts or scenes, but the overall shape of the drama—the rise and fall of the action. Dramatic works can have any of a number of possible shapes, but there is an archetypal shape that can be found at the root of many of them. This shape is often called "the dramatic arc." It begins with a situation, often an apparently stable and peaceful one. Then something

Figure 3.1

happens and a conflict arises, which gradually intensifies. (Many recent movies start with the conflict already established and in motion.) The conflict builds in intensity until it reaches a climactic point, when the conflict is usually resolved, and is followed by a *coda* or *dénouement* in which the tensions that led to the climax are dissipated through talk, through comedy, or through some ritual action or ceremony, such as a wedding. If this arc were to be represented by a linear graph on paper, it would start low and rise gently, then more steeply, until it reaches a high point (the climax) and then subsides. Almost all melodramas—in fact most theatrical works in any genre—share the dramatic arc (see figure 3.1).

As an example of the dramatic arc in a musical, consider the libretto of *Oklahoma!*, which is divided into two acts of three scenes each. The first scene of Act One begins quietly and introduces us to all of the major characters and to the romantic triangles that constitute the main plot and the subplot (Curly-Laurey-Jud and Will-Ado Annie-Ali Hakim). Scene Two intensifies the rivalry between Curly and Jud with a threat of violence, and in Scene Three the dream ballet again raises the specter of violence, ending the act with the question of how the main plot will resolve. Act Two, Scene One intensifies the tensions of both main plot and subplot during the bidding over the picnic baskets, and Scene Two, which takes place a few moments later, resolves both triangles happily. Emotionally, this is the climax of the show, but there is a last flareup of violence in Scene Three as Jud attacks Curly and is killed. This grim situation is wrapped up in a dénouement that takes no more than two pages of dialogue, and the show is over.

Just as the course of the drama usually describes an arc, so does the path of each of its main characters. Generally each character has a problem, need, or desire which that character tries to resolve, and by the end the problem is resolved in some way, happily or not. In the course of writing a play or movie, writers often plot their characters' arcs to make sure that they progress believably and interestingly. Actors preparing to play a role often do the same thing.

In a good play, because of the experiences they undergo, the leading characters learn or change in some way. In *Oklahoma!*, Curly and Laurey learn to express their feelings for each other; in *Guys and Dolls*, Nathan Detroit and Sky

Masterson finally allow themselves to be dragged to the altar; and in *A Little Night Music*, Frederick Egerman gives up his fantasies of life with a young wife. As always, there are exceptions to this principle, such as in farce, which is primarily comedy about situation rather than character; people's personalities rarely change during the course of a farce. Nevertheless, even in *A Funny Thing Happened on the Way to the Forum*, the finale makes a point of enumerating the circumstances that have changed for the major characters: Pseudolus becomes a free citizen, and most of the other characters' lives have changed as well.

Drama is a form of entertainment, and as such it needs variety. Tragedy in particular seems to need comedy. Some writers believe that comedy also needs tragedy, but others disagree, and certainly there have been many comedies with few or no serious moments. As far as we know, the Greeks did not combine tragedy and comedy in a single play, but traditionally each tragic trilogy was followed by a comic "satyr play." Certainly there is no play by Shakespeare, no matter how serious, that doesn't have comic scenes in it. The classic example is the porter scene in *Macbeth*, which takes place immediately after Macbeth's murder of the king. Similar scenes can be found in *Hamlet* and all the other tragedies, and it is hard to think of any great drama that does not contain some comedy.

Another point to keep in mind is that theater is illusion. It can deal with the past, the present, or the future, or it can be a total fantasy, but it does not show real life—even when it pretends to. Most things that occur or are said in everyday life would be excruciatingly boring on stage. Drama, like all the arts, demands a selection from the infinite possibilities of past, present, future, and the imagination. From these possibilities a dramatist chooses a small number of interconnected actions and conflicts—enough to fill a few minutes or a few hours—and using a particular style, viewpoint, and thematic spine, constructs a simulation of life, in order to evoke an emotional response from an audience. Because this is only an illusion, the dramatist can and must take liberties with the raw materials of daily life and historical fact.

THE NEEDS OF THE LIBRETTO

People who have written both plays and musicals usually agree that it is much more difficult to write a good libretto than a good play. The musical theater libretto is a genre with its own requirements, many of which are the same as those of a straight play, but some of which are quite different. The most important differences include the following:

In a musical, at least some of the high points of the drama are set as musical high points. If they are not, the story has not been truly conceived as musical theater, but rather as a play with songs.

A good libretto will necessarily look and feel incomplete without songs; it *needs* songs to make it a dramatic whole. It must contain all the elements that make good drama, but paradoxically, in order to work as part of a musical, it must *not* work on its own. If it works just as well without musical numbers, it may be a good play, but it's not a good musical.

A good libretto is extremely concise. While we will discuss playing length later in this chapter, we can say as a general rule that songs and dances take time, and therefore dialogue cannot. For a show to have an approximately equal balance of songs and dialogue, in an overall playing time of approximately two hours, the writers must leave room for approximately one hour's worth of musical numbers (including songs, dances, reprises, an overture, bows, and so on). Unlike the dialogue in a play, which can be witty and stylish in the hands of a writer like Congreve, Sheridan, Wilde, Bernard Shaw, Coward, or Stoppard, dialogue in a musical is usually concise and utilitarian; the librettist must be able and willing to prune away anything that does not advance the plot, that is not in some way essential to the structure of the story. A rare exception is *The Fantasticks*, in which the spoken portions, especially El Gallo's monologues, attempt to evoke a poetic sensibility. Even here, though, it should be noted that the plot of *The Fantasticks* is extremely simple, which allowed Tom Jones, the librettist and lyricist, room for poetic stylization and ornamentation.

Because the libretto must be relatively short, there is very little time to develop complexity of character, at least in the dialogue. The nuances and shades of the characters' personalities must be developed primarily in the songs.

There is usually a seamless flow between spoken and sung lines, at least in shows written since *Oklahoma!* This is true even in shows that are almost entirely sung. This principle follows from the basic premise of musical theater as a hybrid medium. Clearly it is easier for an audience to get involved in the story, to accept the dual conventions of singing and speaking, if there are no jarring breaks between one mode of discourse and the other.

The exception to this principle is the concept musical, which makes use of what are often called Brechtian methods, from the "alienation" technique used by Bertolt Brecht in the writing and staging of such plays as *The Caucasian Chalk Circle* and *The Good Woman of Sichuan*. In these plays and in concept musicals, which are their descendants in certain ways, no attempt is made to mediate between the different styles of speaking and singing, or between other theatrical styles. Such shows as *Company* and *Chicago* make strong distinctions between dialogue scenes and musical numbers. But there is always a danger that this technique will accomplish precisely what Brecht claimed he wanted, the alienation of the audience. While some plays using this approach have worked with general audiences, very few musicals have. Both *Company* and *Chicago* were only moderately successful in their original productions.

The two most common contributions a librettist makes to the score of a musical are ideas or suggestions for songs, and specific lines of dialogue that the lyricist appropriates for songs. Librettists must be willing, even eager, to take a back seat and let some of their best lines and ideas go into songs and dances. Although lyricists often joke about having "stolen" some of their best lines from their librettists, experienced librettists expect this appropriation: They understand that in a musical, songs are the focal points of interest and excitement, and so it is theatrically right for the songs to have the funniest jokes or the most potent imagery.

An example of this occurs at the end of Act One, Scene Two of *West Side Story*, with Tony's song "Something's Coming." Many theater professionals consider the lyric of this song to be one of Stephen Sondheim's most impressive achievements. It is interesting, it tells us important things about the male protagonist, and it sets up a crucial element of the entire story—and yet it is about almost nothing; all it says is that Tony wants to find something new in his life. But as impressive as the lyric is, Sondheim has been the first to admit that the idea of the song and much of the lyric came from the libretto by Arthur Laurents, who had written a speech for Tony that included such lines as "it may be around the corner, whistling down the river . . . who knows?" And Laurents was happy to have his lines transformed; he has said that the song made a stronger impression than any speech could.

One other practical point should be mentioned here. Librettists have always been the odd people out in the creative team, the unknown soldiers. Their names are rarely known or remembered by the public, and they are often ignored by critics. They are usually blamed if a show does not work, and almost never given the credit if it does. Although they often are responsible for the show's structure, and sometimes its concept and tone as well, they make no more money, and sometimes less, than the lyricist or the composer. In short, librettists must be conversant with the principles and conventions of musical theater; skillful with language and structure; ruthless in cutting their own work; and willing to let their best ideas be taken and used by their collaborators. On top of all that, they must have a thick skin, and no need to share the spotlight or the glory.

A GOLDEN RULE (AND ITS COROLLARIES)

Musical theater has traditionally been not only more of an auditory genre but also more of a *visual* genre, more of a spectacle, than straight theater. With its emphasis on movement and song rather than on dialogue, the musical seems to require that the audience see and hear scenes and actions for itself, rather than hear them described.

Why should this be the case? We can speculate that tradition has played a part. Musical theater arose from a convergence of drama, variety show, ballet, and opera, and people got used to seeing a lot of things happen on stage. Another reason might be that in musical theater, songs are heightened speech, and they seem to require larger and more spectacular actions to balance them.

Thus a golden rule of musical theater writing is: *Don't tell them, show them.* This rule has a number of significant corollaries, which seem to be essential aspects of every good musical. There are exceptions to all of them, but they are valid enough to be taken as working rules for any book show.

1. *The audience gets to know the main characters as early as possible.* As discussed in chapter 2, musical theater is about feeling, and in order for us to care about and root for the main characters, we need to get to know them and like them quickly. The sooner we get to know them, the quicker we start to see things as they do.

Imagine that *West Side Story* began not with the Jets and the Sharks, but with a day in the life of Officer Krupke, a hard-working middle-aged New York cop. He has a wife, three kids, and a job in which he makes little money and gets little respect while risking his life every day, trying to catch criminals and protect innocent people; despite all this, he still manages to keep a sense of humor. Imagine how we would feel if we followed Officer Krupke, at the end of a busy day of frustration and danger, as he attempted to break up a fight between two gangs of teenage hoodlums. We probably would not feel the same way about Riff, Bernardo, or their companions. As with many situations in life, our sympathies in theater depend upon our point of view.

The audience is exposed as soon as possible not only to the main characters, but also to the central theme of the show and the central concept, if there is one. Sometimes this happens in the opening number, such as the "Runyonland" sequence from *Guys and Dolls* or "Wilkommen" from *Cabaret*; sometimes it happens in the first scene, as in *My Fair Lady* or *Gypsy*; and sometimes it happens a little more gradually, as in *Kiss Me, Kate* or *The King and I*. But it always happens early in the show.

2. *The protagonists are who and what they seem to be.* This may seem like a minor point—but consider how often, in mysteries, melodramas, action movies, and psychological dramas, characters turn out to have been concealing who or what they truly are. Yet there are no examples of such deception in the main characters of the best musicals, and for good reason. There are few things an audience

hates worse than to find that it has given its sympathy and allegiance to the wrong people, and that it needs to switch its affections in midstream.

This does not mean, of course, that characters can't grow or change during the show; character growth is an important aspect of most drama. But we must *like* our protagonists from the beginning, and in order to like them we must get to know them. If we don't, then we are watching a show in which we do not care what happens to the main characters—and if emotion is crucial to musical theater, then an audience that doesn't care about the main characters is disastrous. There has never been a successful musical with a protagonist for whom the audience did not care in some fundamental way.

3. In musical theater, *songs carry a presumption of sincerity*. When characters sing, we ordinarily assume that they are being honest, consciously at least. (They may reveal things to us of which they are unaware.) This comes from the convention that people sing when their emotions are too strong for speech. There is nothing new about this convention; there are clear parallels in Greek and Roman drama and in Shakespeare. When Hamlet tells us how he feels about his uncle in asides and soliloquies, we accept that he is telling us the unvarnished truth as he sees it, just as we do when Billy Bigelow sings to us.

For this reason, explicit subtext in songs is relatively rare. This is another difference between musical theater and opera; in opera, where usually everything is sung, subtext is quite common. The only prominent exception to this is in the work of Stephen Sondheim, who has said that he prizes subtext. It can be found in a number of his songs, such as "Small World" from *Gypsy* and "In Buddy's Eyes" from *Follies*.

There are some other exceptions to this principle. Sometimes a song is clearly sarcastic. There is also a certain type of presentational number in which a character puts on a false or public face to manipulate another person or a group. In this case we accept that, for specific reasons, the character is not "speaking from the heart" as in other songs, but simply appearing to do so, as people often do in real life. This category includes "Ya Got Trouble" from *The Music Man* and "We Both Reached for the Gun" from *Chicago*. "Tevye's Dream" from *Fiddler on the Roof*, a long number utilizing the entire cast, is one big lie, concocted by Tevye for the purpose of getting Golde to accept a poor tailor for a son-in-law.

4. *Important events take place onstage*. This is a logical consequence of all the above. However, veteran theatergoers may immediately think of a prominent exception to this rule: "You Did It," the first song in the second act of *My Fair Lady*. This is a major number with soloists and chorus, describing how Eliza had successfully fooled Professor Karpathy after the final scene of Act One. In the original play these events were shown onstage, but it is not difficult to see

why Lerner and Loewe chose to have such an important song describe offstage events. A scene from the original play in which Professor Karpathy mistakenly announces to a group of minor characters that Eliza is a Hungarian princess is not as useful to the story as the scene from the musical in which Higgins and Pickering recount the story and crow over it while ignoring Eliza, which triggers the events of the rest of the show. In addition, while Karpathy's announcement might have made an amusing song, he is still a minor character, while Higgins is the male leading role. Finally, to end the first act before Karpathy's announcement, so that the audience is not sure whether or not Eliza's masquerade will succeed, gives the show a much more suspenseful first-act curtain. Clearly it made more sense to replace the rest of the scene with "You Did It," an extremely effective and delightful number.

5. *The principle of opposition applies to musical theater*, even more than to other forms of drama. In order to induce the audience to have a certain feeling about a character, it is necessary for the character to evince the opposite feeling. Take the common situation of a character feeling unhappy about something. If she rambles on in self-pity about her situation, the result is usually comedy—for instance, "Adelaide's Lament" from *Guys and Dolls*, or "You Can Always Count on Me" from *City of Angels*. On the other hand, when a character is upbeat and optimistic despite adverse circumstances, we feel all the more strongly for her. A perfect example of this is the first song in *Annie*, "Maybe." Although Annie is a lonely and mistreated orphan, she does not moan about how miserable her life is. Instead she is optimistic and chooses to focus on the possibility of finding her parents. Immediately we feel sympathy for her pluck and bravery despite the odds. Many theatergoers have a violent dislike for Fantine's big self-pitying number, "I Dreamed a Dream," in *Les Misérables*. If the show works despite this, it is only because she is a comparatively minor character. When characters go on and on about how miserable their lives are, the audience often loses sympathy for them or starts to laugh. Good librettists don't tell us how hard the character's life is, they show us.

We will encounter other applications of the principle of opposition throughout this book. For example, good lyricists never use a trite phrase like "believe me" to indicate that the character singing is sincere. They know that a person who says such things in real life is usually lying. They similarly avoid such empty phrases as "I'm certain" or "I know it"; as in real life, these phrases almost always convey the opposite meaning.

6. *Narrative momentum steadily builds toward a resolution.* As important as constant forward drive is in most drama, some writers believe that it is even more important in musicals. (Tom Jones refers to it as both "thrust" and "pace.") The many interruptions of the narrative by song and dance, which can "stop the

show" and make for exciting moments, can also weaken the momentum and the cumulative effect of the drama. Thus all musical numbers must advance the story in some way. (We will discuss the *ways* in which songs can advance the story later in this chapter.) In addition, almost every scene ends with something important unresolved, leading the audience toward the next plot point.

7. *Good musicals don't preach to the audience.* As discussed in chapter 2, the essence of musical theater is not the advancement of a thesis or argument, but the conveying of an emotional experience, an illumination of some aspect of human life—both through behavior and through songs that amplify the emotions inherent in that behavior. None of the great musicals has had an explicit message. Such implicit messages as they may have conveyed have tended to be romantic generalities, such as "love will conquer all." To update an old theater maxim: If you have a message, send an e-mail.

In addition, messages tend to date rather quickly. As an example, consider Lieutenant Cable's song from *South Pacific*, "You've Got to Be Carefully Taught." While well meant and unobjectionable in its sentiments, it no longer holds up very well. If it does not drag the show down with its rather explicit sermon against racism, that is because it helps to illuminate Cable's character and deals with one of the main themes of the show, in both the Cable-Liat subplot and the primary de Becque-Nellie plot as well.

FORMS AND FORMATS

The outer shapes of drama have changed considerably over the centuries. Classical Greek tragedies were each in a single act, and usually observed "the unities," of which the most important were those of time and place. The events of a play took place over a period as long as the duration of the play itself—that is, in "real time"—and in a single location. On the other hand, Elizabethan dramas such as Shakespeare's had no such unities; in five acts, they ranged through many places and sometimes covered significant gaps of time. The nineteenth and twentieth centuries produced many examples of what came to be called the "well-made play," which was usually in three acts—although occasionally in four or even five—with a single time period and locale for each act or for the entire play.

The vast majority of musicals have been in two acts. The first act sets up the situation and the conflict, and the second act resolves it. Most of these musicals have numerous scenes and locations in each act, like Shakespeare's plays. The most prominent exception is *A Funny Thing Happened on the Way to the Forum* which, like its classical models, takes place in a single location and in the same amount of time as the show itself, not counting the intermission.

Recent decades have seen more flexibility in the outer shapes of musicals. There have been a number of shows in a single act, such as *1776*, *Follies*, and *A Chorus Line*. There have even been a few in three acts, such as *The Most Happy Fella*, *Anyone Can Whistle*, and *The Apple Tree*. Some musicals without an intermission observe the unities of time and place, like *Follies* and *A Chorus Line*. But no matter how many acts they have, many musicals take place in a relatively short span of time, illustrating the principle of narrative momentum (corollary #6 above). If there is a jump in time between scenes, the scenes on either side of the break tend to cluster into short spans of time. For example, the first three scenes of *My Fair Lady* take place within a twenty-four-hour period. Then there is a jump of three days before Scene Four, to establish that Eliza has moved in with Higgins and Pickering. The next scene, which begins later the same day, condenses time by showing Eliza's gradual mastery of the English language over a period of several months, in a series of short exchanges. The following three scenes also take place within a single day, and after another jump of six weeks to establish that Eliza has improved enough to go to the ball, the final three scenes of the act and the entire second act all take place within a single day.

One of the crucial questions for any musical is where the act break or breaks will occur. If musical theater is primarily about emotion, it makes practical sense to leave the audience at a moment when it most desires to find out what happens next. In the two-act form, the most common place for such a break is a little more than halfway through the show, at a moment of suspense or uncertainty, such as: Will Eliza succeed at the ball and win the bet for Higgins? Will Harry Beaton break the spell and destroy Brigadoon? How will Pseudolus win his freedom when he is about to be put to death? Consider the act break of *Carousel*. It is not at the most *exciting* point in the story—the bungled robbery and Billy's death—but earlier, at the most *suspenseful* point, when everyone is leaving for the clambake and we are worried about what Billy will do next.

The lack of an effective act break may have caused problems for even such a well-crafted show as *Pal Joey*. Act One of *Pal Joey* ends with a dream scene in which Joey, an untalented and unscrupulous performer who has become the boyfriend of a wealthy woman, imagines the fancy new nightclub he will soon be running. Until this point there has been no serious threat to Joey's rise, and because the audience does not really care about him one way or the other, there is little suspense about what will happen next. The lack of a sufficiently suspenseful act break may have hurt the popularity of this widely respected but infrequently revived show.

While questions such as the one in *Oklahoma!*, over whether Laurey will go to the dance with Curly or with Jud, may no longer be compelling enough, the uncertainty at the act break does not need to be a matter of life and death. The crucial element is not suspense in the melodramatic sense, although that can certainly be effective, but the desire of the audience to come back after intermission to see and hear more. As Lehman Engel wrote, even when the audience expects that everything will turn out well, at a successful show it still wants to see *how*. This can come from having enjoyed any number of elements of the show: the dialogue, the score, the performers, the direction, even the choreography. Shows that have managed to bring back their audiences despite having no significant suspense at the act break include *The Fantasticks, Into the Woods*, and *The Producers*.

Choosing a place for an act break, or even whether to have one, can be a difficult decision, and a significant number of shows have moved, removed, or inserted act breaks before opening. When the creators of a show can find no logical point for an act break, or when they feel that an intermission would interrupt the narrative momentum and flow, they sometimes omit an act break altogether.

A one-act play or musical is not a full-length show without an intermission, but rather a half-length show that usually runs for an hour or less. A one-act show does not constitute a full evening of theater, and usually it is coupled with other one-acts. There are many such anthology "evenings" of one-act plays, although anthology musicals are rare.

Musicals for children, which are usually in one act, have traditionally been in a separate category altogether. The big Disney musicals which began to appear on Broadway during the 1990s are full-length shows in two acts; most of the children attending them are accompanied by adults; and many lines of the libretto and lyrics are aimed more toward the adults than toward the children. Thus these Disney musicals are really "family" shows for a general audience of children and adults, like *Peter Pan* or *Annie*, rather than true children's musicals.

MODES OF STORYTELLING

There are many different ways for a musical to convey its story. The most important include speech, rhyme, recitative, song, and movement.

Speech is a generic term for the "realistic" prose dialogue that forms a significant part of most traditional musicals, and appears to some extent in almost every musical, even the so-called through-composed shows. Speech is also used occasionally for narration, as in parts of *Into the Woods, Sunset Boulevard*, and the prologue of *Beauty and the Beast*. *Rhyme* is of course a common element of most song lyrics, and perhaps for that reason it rarely appears in the spoken sections of musicals. We will discuss rhyme and other aspects of lyrics in chapter 5.

Recitative (pronounced *reh-chih-tah-TEEV* or *ray-see-tah-TEEV*) is a term from opera, denoting passages that are sung but are not part of a song. These are the prose "talky" sections of many operas that precede or follow the featured songs. Recitative passages have only simple chords as accompaniment, often played by a single keyboard instrument, as in Mozart's Italian operas. In these operas, recitative fulfills the same function that speech does in operas with spoken dialogue. Nevertheless, recitative can also be found as a transition between dialogue and song in some operas (such as Bizet's *Carmen* in its original form). It can also be found in many operettas and musicals, although it is usually not identified as such. In the middle of "A Boy Like That" from *West Side Story*, Maria's lines beginning "Oh no, Anita, no" are recitative; and in "Tevye's Dream" from *Fiddler on the Roof*, so are the ensemble's lines beginning "Shah! Shah! Look!" and Fruma-Sarah's lines beginning "If Tzeitel marries Lazar Wolf." The transitional function of recitative is analogous to, and sometimes indistinguishable from, the verse or interlude sections of songs, although ordinarily the latter are rhymed and recitative is not.

Song, the union of lyrics and music, is discussed in great detail throughout this book. *Movement* is a general term that covers any kind of preset or coordinated gestures or steps made by the performers; these include pantomime, stylized "realistic" gestures, coordinated or synchronized group movements, and dance. Virtually every style of dance may be found in musical theater, but the most common have been ballet, tap dance, and the many varieties of "jazz dance."

Movement is something of an anomaly in the list of storytelling modes, however. Musicals have run the gamut from shows driven by or largely told through dance (such as *West Side Story* and *A Chorus Line*) to shows with virtually no dance or choreographed movement (such as *Passion* and *You're a Good Man, Charlie Brown*). Thus while a traditional musical must have songs, it does not necessarily need movement. As mentioned in chapter 1, in recent years there have been a few shows with no singing at all, *only* movement—mostly choreographic revues like *Dancin'* and *Fosse*, but also *Contact*, which could be considered a book show. But these will probably remain rare exceptions. For most people, musical theater remains essentially a genre with song.

Movement, of course, can be as dramatic as any other aspect of a musical. Anyone who has seen the original choreography of Michael Kidd for *Guys and Dolls*, of Onna White for *The Music Man*, of Jerome Robbins for *West Side Story*, of Ron Field for *Cabaret*, or of Michael Bennett for *Follies*, knows that theater dance at its best has been an integral part of musical theater storytelling, not just an exciting diversion. Jerome Robbins in particular was a master of dance as storytelling. Some of the most moving moments in *West Side Story* occur during the "Somewhere" ballet sequence, one of the peak achievements of musical theater as a combination of drama, song, and dance. Arthur Laurents has attributed Robbins's supremacy to the fact that he never choreographed a musical number simply as a dance, but always as a dance *about something*. Those rare choreographers who understand not only movement but also drama can make movement dramatic, creating dances that help to tell the story rather than interrupt it. Such choreographers make a show much more unified and exciting—but they also make the fate of the show utterly dependent on their own taste and ability. There are many stories in Broadway history of shows that started out with promise, only to be sabotaged by unsympathetic or inappropriate staging.

THEATRICAL STYLES

Just as musical theater writers have a wide range of storytelling modes at their disposal, so they also have a wide range of dramatic styles or types of drama available. These include narration, monologue, tragedy, various kinds of melodrama and comedy, farce, mystery, satire, and parody, as well as other, less venerable styles such as "camp." (This is the nostalgic recreation of another period and its style. Often it exaggerates that style as part of a satirical approach, making fun of the story and characters.)

Some theatrical styles are more difficult than others to make work in the genre of musical theater—for instance, farce, which is primarily about situation rather than character. In farce, cardboard characters frantically rush around and go nowhere, like cogs in a machine, rarely undergoing any change. Farce is rarely about emotion, except perhaps manic desperation, and is usually inappropriate for musicalization. Nevertheless, *A Funny Thing Happened on the Way to the Forum* proves that it is possible for a superlative farce to be a superlative musical. Unlike the complicated farces of Feydeau, for instance, *Forum* uses classic comic types, each one recognizable and distinct—the henpecked husband, the braggart warrior, and so on—and a convoluted but essentially simple plot. In addition, *Forum*'s protagonist has an objective which is more compelling than the conventional farcical drives of money and sex (although he pursues those goals as well).

For similar reasons, topical satire, camp, and mysteries do not usually adapt well to musicalization either, and the few exceptions that have succeeded did so for very specific reasons. As successful as *Finian's Rainbow* was in its initial run, and despite a brilliant score, its political satire has not held up well. On the other hand, the gentler and more generic satire of *How to Succeed in Business without Really Trying* still seems to strike a chord. Similarly, there have been many camp musicals that have died on or off Broadway. *Little Shop of Horrors* succeeded where all the others failed, not only because it is far better written, but because it has genuine affection for its leads and their romance. As for mysteries, they are inherently unsuitable for musicalization because the essence of a mystery story is that no character, except perhaps a detective, is really what they seem. This means that any character for whom the audience has started to care could turn out to be quite a different person. In addition, like farces, mysteries tend to be more about situation than about character. The only moderately successful exception, *The Mystery of Edwin Drood*, made the central mystery story into a show-within-a-show, in which the mystery itself was less important than the hijinks of the performers. Similarly *Something's Afoot*, which failed on Broadway but has been moderately successful elsewhere, treated its mystery as a subject for comedy.

Nevertheless, it might still be possible for gifted writers to make a mystery work as a musical. One way would be for the detective to be the protagonist, with the focus on his (or her) personal life. Even suspects could be protagonists, if they were clearly indicated as protagonists and remained as likeable as they initially appeared to be. Even if they turned out to be concealing something minor, such a show could create audience empathy and genuine emotion. But of course that would effectively remove the protagonists from the status of potential suspects.

It should be clear that certain other types of plots will not work well as musicals, such as any story that is told backwards. Although this device may be interesting as a conceit—and Tom Stoppard and Harold Pinter have both made theatrically effective use of it—it seems to be inherently unsuitable for a musical, because we only get to know the characters indirectly and "in reverse." Thus the show *Merrily We Roll Along*, despite a strong score by Stephen Sondheim, has never truly worked in any of its versions. The main character is first seen as a phony, uninteresting middle-aged man, and no one cares that by the end of the show he is an idealistic eighteen-year-old. (The original play on which *Merrily* was based was one of Kaufman and Hart's few flops.)

This problem could be circumvented, of course, in a show about someone who starts as a sinner and ends up a saint. The question remains, however, why that story should be told in reverse. What dimension or emotional power does this theatrical conceit add to a musical? Similarly, to make a musical from any story in which the characters are all revealed at the end to be puppets of some higher power, or symbols, or part of someone's dream, is asking for trouble.

BALANCING THE ELEMENTS

Even after a story has been chosen and a libretto begun, writers still have many choices to make. Decisions about which parts of the story should be set as songs, which as dance, and which as dialogue can be extremely difficult, and are often subject to change. Even when each of these moments works separately, they may not work together in the show if the balance is too heavily tipped in one direction or another.

To a considerable extent, such decisions depend on the overall style of the show. As examples of the range of possibilities within commercial musical theater, we might mention on the one hand *Jesus Christ Superstar* or *Les Misérables*, which are almost entirely sung, and on the other *The King and I* or *A Funny Thing Happened on the Way to the Forum*, in which the songs take less than half of the running time. Most musicals have fourteen or fifteen songs in their score. But Stephen Sondheim's score for *Pacific Overtures* has eleven songs, while his score for *Sweeney Todd* (his next show) has twenty-two.

In the Introduction we mentioned that one of the primary differences between opera and musical theater is that in a musical, most high points are either sung or danced, but in an opera, *all* the high points are musicalized. This is not a hard and fast rule, of course, but even in the case of "number" operas with spoken dialogue, such as Mozart's *Magic Flute* or Bizet's *Carmen*, virtually all the dramatic high points are musical ones.

In musical theater, on the other hand, there are several famous instances where climactic points of a show are *not* sung. One comes from the last scene of *West Side Story*, when Maria picks up the gun that has killed Tony and confronts the gangs and the police. Surely here, if anywhere, is a place for the female lead to have a climactic song, an "eleven o'clock number," but instead Maria has a long speech. The reason is simply that the songwriters were unable to come up with a song that satisfied them. In this case they were fortunate, because at the end of a powerful evening of musical theater, this bare speech is enough to work without any music at all.

Another prominent example of a climactic point that is not musicalized comes from the end of Act One of *Camelot*. King Arthur has realized that his wife and his best friend are in love with each other. In this case, the writers decided that to have Arthur sing this moment would be too operatic, and that since Arthur was being played by Richard Burton, a star with one of the great speaking voices of the century, a soliloquy with underscoring could be as effective as a song.

A third well-known example comes from halfway through *A Chorus Line*. One of the most powerful parts of the show is when Paul, the shy dancer near the end of the line, tells his story of working in drag shows and trying to find his identity as a man. It is also the only scene that is not musicalized at all; it does not even have underscoring, except at the very beginning and end of the scene. As in *West Side Story*, the songwriters always intended to replace it with a song, but apparently could not find one that was more effective than the simple monologue. Part of this scene's emotional power, of course, is that it follows two long and overwhelming musical numbers (the immense "Montage" and Cassie's solo "The Music and the Mirror") and then is followed by the long and overwhelming "One." The dramatic contrast of Paul's quiet confession makes it all the more poignant and memorable.

As always, there are no absolutely right or wrong solutions. But it can be taken as a general rule that if the most dramatic moments in the main characters' lives are not musicalized, there is probably something wrong with the storytelling. Many shows in recent decades are all or mostly sung, and they may seem to avoid this problem, but they only sidestep it: Even if everything is sung, there still has to be a *reason* for the characters to sing—a reason that some rock operas and "poperettas" have not provided. If characters in a musical don't sing to express the most emotionally charged moments, then why would they ever sing?

Translating the Story into a Script

First, writers have an idea for a show. How do they translate their idea into a script? Whether the libretto is an adaptation of material from another medium, or is based on an original idea, it requires the fundamental elements of good drama and good musical theater that have already been discussed.

Often, if not always, it requires one more thing, which we call the theme, spine, or core of the show. Whatever it is called, it is the very essence of the story, and it has to be kept in mind by the writers throughout the entire process of writing, rewriting, rehearsing, and performance. Almost anything else in the show—any character or song or detail—can change except the central core, the reason why the writers chose to write that particular show in the first place. Without a central core that the creators agree upon, a show can change direction or lose its "soul" during development, rehearsals, or previews, and turn into something very different from the show they set out to write. To paraphrase a simile of William Goldman's, it is like a sailing ship heading north that, because of unforeseen problems, has to keep tacking into the wind and altering its course, until eventually it finds itself going south instead. Goldman provides several examples of shows that got lost on their way to Broadway in his book *The Season*.

Experienced directors and writers know how essential it is to grasp this spine or core of the work. Goldman's brother James, author of *The Lion in Winter* and librettist of *Follies*, would tape a piece of paper on which he had written "What is the show about?" to the bathroom mirror of his hotel room when out of town, so that he would not lose track of the show's spine in the midst of rewriting. Similarly, from the beginning of their collaboration on *Fiddler on the Roof*, director Jerome Robbins badgered the writers with one question: "What is this show about?" and refused to accept answers that simply summarized the plot, like "It's about this dairyman who has five daughters." Months later, one of them was finally able to verbalize that the theme of the work was the changing of old traditions—a center around which the entire show and production coalesced. And when director Robert Lewis read an early version of *Brigadoon*, he asked librettist Alan Jay Lerner if he realized what he had written. Lerner, confused, started to summarize the plot—it was about a town in Scotland that comes to life every one hundred years, and so on. Not at all, Lewis replied; it was the story of a romantic who is searching and a cynic who has given up, and at the end the cynic is proved wrong. Lerner immediately realized that Lewis was right—and after months of problems with the script, he was able to finish the final draft in a week.

Adaptations of Plays and Movies

In general, a dramatic work such as a play or movie is the least problematic source for a musical—which probably accounts for the large percentage of musicals that are based on plays and movies. Writers who adapt a good play or movie have had many problems solved for them; they already have characters, conflict, action that develops and builds through varied scenes, and so on. But since a musical is different from either a play or a movie, considerable recasting of the material is always necessary. When writers adapt a property for musical theater from another genre, almost anything about the original may have to be discarded. The only exception is the central core of the work, the theme or feeling or character that made the writers want to adapt it in the first place.

Movies often resemble musicals in that they tend to have many short scenes rather than a few long ones. So in musical adaptations of movies, much of the work of dividing the action into scenes will have been done already. The major problems in movie adaptations are usually due to the practical and financial difficulties of putting some of the movie's locations, set pieces, or crowd scenes on the stage.

Play adaptations have the opposite problem. Especially in the last two centuries, plays have tended to use small casts and a small number of sets and scenes, so the challenge is usually to "open up the play," to intersperse the original scenes with others that have contrasting locales and characters. A perfect example of this can be found in *My Fair Lady*. The libretto contains many scenes that do not appear in *Pygmalion*, the original play—but most of them do appear in the *movie* of *Pygmalion*, which appeared eighteen years before *My Fair Lady*. (Incidentally, the published libretto bears a preface by Alan Jay Lerner in which he justifies his having deviated from George Bernard Shaw's original ending for the play by saying, "Shaw and Heaven forgive me!—I am not certain he is right." Perhaps not, but Lerner fails to mention that like many other aspects of *My Fair Lady*, the show's ending copies the ending of the movie, the screenplay of which is credited to Shaw himself.)

By now, writers have many successful adaptations of dramatic material from which they can learn. They can compare *Green Grow the Lilacs* with *Oklahoma!*, *Liliom* with *Carousel*, *They Knew What They Wanted* with *The Most Happy Fella*, and *Romeo and Juliet* with *West Side Story*. For an illustration of the adaptation of a dramatic work into a musical libretto, let's take a look at the adaptation of the movie *Sommarnattens Leende (Smiles of a Summer Night)*, written and directed by Ingmar Bergman, into the musical *A Little Night Music* by librettist Hugh Wheeler, songwriter Stephen Sondheim, and director Harold Prince. We will compare synopses of the movie and the musical. Discrepancies in spelling and nomenclature in these synopses follow the discrepancies between the film and the musical.

SMILES OF A SUMMER NIGHT

Smiles of a Summer Night is "a romantic comedy" that takes place in a small town in Sweden in 1901. Neither the movie nor Bergman's published screenplay is divided into acts or scenes, so the divisions of the action below are for convenience. Naturally this summary omits many details, both realistic and atmospheric, supplied by the camera.

On an afternoon in late spring, the middle-aged attorney Fredrik Egerman closes his books and leaves his office happily. He walks into a photography studio and picks up an order, pictures of his eighteen-year-old wife Anne. Then he passes a poster advertising the appearance of a theater company that night, starring the famous actress Desirée Armfeldt. Egerman hesitates, then buys two tickets.

At home he greets his maid Petra, who is Anne's age. Henrik, Fredrik's son from his first marriage, a severe nineteen-year-old who has just graduated from a seminary, is reading about Martin Luther to Anne, who greets her husband happily.

Fredrik and Henrik are more formal with each other. Fredrik shows Anne the tickets to the play, and they go to their room for a nap until curtain time. Henrik accuses Petra, who has come in to clear the tea service, of walking in a deliberately provocative manner. He suddenly kisses her; she slaps him, then leaves, exaggerating her movements. Henrik angrily starts to play the piano. Anne enters and asks him to be quiet, because his father is already asleep. She returns to the bedroom and lies down next to Fredrik. He seems to be having an erotic dream, and caressing her in his sleep, he calls her Desirée. She pulls away and begins to cry.

At the theater, Fredrik and Anne sit in a box close to the stage; the play, a drawing-room comedy, is in progress. Desirée Armfeldt makes a grand entrance. Anne examines her carefully. She insists that Desirée is looking up at their box. During Desirée's first scene, full of artful banter about love and sex, Anne starts to cry and demands to be taken home.

When Petra opens the front door for them she looks disheveled. She puts Anne to bed, while Fredrik is surprised to find Henrik also disheveled and blushing. Fredrik congratulates him: It is spring and Henrik is young, so it is appropriate that he should celebrate. Henrik says he has sinned, and it was a failure. Petra comes out of the bedroom and tells Fredrik that Anne wants to say good night. Anne asks Fredrik if he really loves her, then starts to reminisce about how when she was a little girl he would tell her stories until she fell asleep, and she called him "Uncle Fredrik." She speculates that Desirée must be very old, kisses him good night, and says that one day she will be truly his, but he must be patient with her. Fredrik says he will sit up for a while. He tiptoes out of the bedroom.

Backstage at the theater, the play has just ended as Fredrik enters. Desirée is happy to see him, and leads him to her dressing room. She is surprised that Fredrik, whom she knew as a dedicated womanizer after his first wife died, has become a devoted husband, and senses that he is unhappy. He mentions his erotic dream. He confesses that Anne is still a virgin; he is in love with her, but frustrated that she treats him like a father. He asks Desirée to help him: to tell him it's hopeless with Anne, or anything else. Desirée invites him to her rooms for a drink. As they walk through the dark streets, Fredrik steps into a deep puddle and gets soaking wet.

At Desirée's house, she has dressed Fredrik in a nightshirt and robe while his own clothes dry. He asks whose clothes he is wearing, and she tells him about her lover, a handsome soldier. Her four-year-old son walks in and she calls him Fredrik. Fredrik Egerman is astounded, but Desirée assures him that the child is named after Fredrik the Great. Fredrik says she is unfit to have a child, and she slaps him, telling him to go. He apologizes. He reminds her that it was she who broke off their relationship; she says that she was only a playmate for him, and that he cheated on her, which he admits. Their discussion again grows acrimonious, and Fredrik is about to leave when there is a pounding at the door. It is Desirée's lover, Count Carl-Magnus Malcolm. She warns Fredrik that he is very jealous and dangerous. Malcolm enters; he has twenty hours' leave, of which nine are for Desirée and five for his wife. He wants to put on his robe, but it seems to be occupied. He is polite but icy with Fredrik. Desirée goes out to see if Fredrik's clothes are dry, and Malcolm mentions his proficiency at dueling, which he demonstrates by throwing a fruit knife at a target. Desirée returns; Fredrik's clothes aren't dry yet, but Malcolm insists that he leave immediately.

The next morning, Desirée goes to see her elderly crippled mother. Mrs. Armfeldt lives in a castle given to her by an old paramour. Desirée tells Mrs. Armfeldt that she has broken off her affair with Count Malcolm, and has a new man in mind. She asks her mother to throw a party, and to invite Malcolm, his wife, and the three Egermans. She admits that she still loves Fredrik.

Count Malcolm, at home, practices his shooting. His wife Charlotte enters, surprised that he is home. Their conversation is frank but cool. She asks him how Desirée was. He tells her about Fredrik in the nightshirt. He adds that they have been invited to Mrs. Armfeldt's estate for the weekend, and that the Egermans will be there. He suggests that she pay her friend Anne Egerman a visit, and tell Anne about her husband's activities.

On the same morning, Anne is in bed. Fredrik goes into his study to work. As young Henrik eats breakfast, Petra the maid passes him and ruffles his hair. He stares, shocked, as she approaches him, unbuttons her blouse, and grabs his hand. He runs into his room and slams the door. Anne asks Petra to brush her hair. They talk about sex and men. When they are finished, Anne is bored and enters Henrik's room. He is reading. She starts to pester him, then demands that he give her his robe, slippers, and pipe because they are disgusting. Then, saying that he has been flirting with Petra, she slaps him. Tears come to his eyes and then hers as well, and she runs out. Then she is delighted to find that Countess Charlotte Malcolm has come to see her. They talk pleasantly until Charlotte mentions Fredrik and reveals his trip to Desirée's house. Anne pretends to know about it already and turns the conversation to Charlotte's husband and his affair with Desirée, which is common knowledge. Suddenly Charlotte begins to cry. She hates her husband and yet still loves him, despite her humiliation. As she calms down, Fredrik enters and tells Anne that they've been invited to Mrs. Armfeldt's estate. At first Anne refuses to go, then changes her mind.

On a sunny afternoon, Mrs. Armfeldt greets the Egermans on the lawn before her castle. Petra unloads the luggage with Mrs. Armfeldt's coachman Frid, a big jovial man who flirts with her. He shows her the rooms for the Egermans, and explains that Henrik's room was built for a king who was dallying with his minister's wife. The minister and his wife stayed in the next room, and when the minister slept, the king would press a knob that would open the wall soundlessly, and slide the bed on which the wife was lying into the king's room. Frid demonstrates the mechanism to Petra.

Desirée comes out of the castle to greet the Egermans. They hear a bang and see the Malcolms coming up the drive in a sputtering automobile. Everyone exchanges polite greetings. Desirée leads Charlotte to her room. Charlotte asks why she has been invited, and Desirée tells her that they should make peace, because they have mutual interests. She has a plan for Charlotte to get her husband back, and for Desirée to get Fredrik.

That night the company sits in formal clothes at dinner, with Mrs. Armfeldt on one side of the long table facing all her guests. Fredrik has been seated next to Charlotte, Count Malcolm next to Desirée, and Henrik next to Anne. All banter gaily and superficially except for Henrik and Anne. Malcolm says that all women can be seduced, while Fredrik asserts that it is the men who are always seduced. Charlotte also disagrees with her husband and bets him that she can seduce Fredrik in less than fifteen minutes. Henrik is disgusted by this talk and smashes his wineglass. When Fredrik reproves him, he angrily retorts that he is ashamed of his father. The others gently ridicule his earnestness, and he suddenly rises and staggers away from the table. Anne begs to be excused and leaves with Petra. The company rises and while the others go outside for coffee, Charlotte follows Fredrik to a window, flirts with him, and kisses him.

Henrik wanders disconsolately through the castle. He is ashamed of himself, calling himself ugly, evil, and stupid. He goes to his room and tries to hang himself with the belt of his robe, which promptly drops him on the floor. He staggers against the wall, and suddenly a bed on which Anne lies sleeping glides into the room. He approaches her and kisses her. She wakes slowly and smiles at him. They confess that they love each other.

In the carriage house, Frid and Petra have been making love as the sky starts to lighten. He tells her that the summer night smiles three times, and this is the first smile, for young lovers. Suddenly Henrik appears with Anne, who embraces him. Fredrik, standing among the trees, sees them. Henrik and Anne climb into the carriage as Petra and Frid load their luggage and harness a horse. Fredrik starts to cry out but stops himself. The carriage drives away.

In her room, Desirée looks out her window at a pavilion by the lake and sees a light within. Fredrik and Charlotte have entered the pavilion; he tells her that Anne has eloped with Henrik, that he knew that they were infatuated but had never minded it, and that now he is devastated. She kisses him and bites his lip. Meanwhile, as Desirée worries that Charlotte will take Fredrik, Count Malcolm enters her room. He is confident that Charlotte is asleep, but Desirée tells him that she is in the pavilion with Fredrik. He rushes out.

Malcolm enters the pavilion and insists that Charlotte leave so that he and Fredrik can play "roulette." Although concerned, she obeys him, and Malcolm pulls out a revolver. If they were to duel, Fredrik would have no chance, so instead they will play Russian roulette. Charlotte meets Desirée on the path and tells her that the men are playing roulette. Malcolm pulls the trigger first. They take turns until Fredrik pulls the trigger a second time.

On the lawn, Desirée and Charlotte hear a shot. Malcolm comes out of the pavilion holding the gun, and when he sees the horror in their eyes he bursts out laughing. He used a blank filled with soot; as a nobleman, he would never risk his life with a lawyer. Desirée rushes into the pavilion. Malcolm calls his wife an unfaithful bitch, like all women. She reminds him of the bet they made at dinner; she has won. She tells him to look at her, that he has never truly looked at her, and that what she wants for having won the bet is him: his promise of faithfulness, for a time at least. He agrees.

It is just before dawn, and Frid and Petra lie on a haystack. He tells her that the summer night is smiling for the second time, for the clowns and fools. Half playing, half fighting, she demands that he marry her, as he said he wanted to.

The clock strikes three. The summer sun is rising. Fredrik sits in a chair in the pavilion as Desirée cleans his face. She puts him on a divan and he falls asleep.

Petra is straddling Frid. She demands that he promise to marry her, and he does. They rise and face the sun, and he says the night has smiled for the third time, for the sad, the sleepless, and the lonely. They walk happily through the grass.

A Little Night Music

The overture is sung by the Quintet (the "Liebeslieders"), five singers who act as a Greek chorus throughout. They start to waltz and are joined by the main characters waltzing in a strange half light, occasionally changing partners.

Act One. Prologue. Madame Armfeldt, playing solitaire, is brought onstage in her wheelchair by her butler Frid, as her granddaughter Fredrika, a grave girl of thirteen, watches. She tells Fredrika that the summer night smiles three times at the follies of people: first at the young, who know nothing; next at the fools, who know too little; and finally at the old, who know too much.

Scene One. The Egerman house. Anne Egerman, who is eighteen and somewhat bored, enters the parlor where her nineteen-year-old stepson Henrik, a divinity student, is playing gloomily on his cello, and teases him. Henrik would like to talk to her about his inner thoughts, but she puts him off. Anne's middle-aged husband

Fredrik enters. Henrik tries to tell him about his examination, but Fredrik puts him off, and shows Anne the theater tickets he has bought. Anne is excited about seeing the famous Desiree Armfeldt. Fredrik follows her into the bedroom and tries to kiss her, but she moves away, chattering excitedly. She asks if she still makes him happy after eleven months, and promises that soon she will be ready for him. Fredrik gives up and decides to take a nap ("Now").

Petra the maid, who is a few years older than Anne, enters the parlor where Henrik is reading and ruffles his hair. He orders her to leave him alone and she walks out, wiggling her hips. When he tells her to stop, she exaggerates it. He lunges at her and starts to kiss and fondle her. She slaps his hand—it's a new blouse—and puts him off. Henrik feels that the whole world is putting him off ("Later"). In the bedroom, while Fredrik sleeps, Anne muses that despite her fear of intimacy, she and Fredrik love each other ("Soon"). She hears Henrik playing his cello loudly and tells him to be quiet, because his father is already asleep. She returns to the bedroom and hears Fredrik, in his sleep, call out to "Desiree."

Young Fredrika Armfeldt is practicing piano scales. Unlike most mothers, hers is an actress. Desiree writes letters to her and Mme. Armfeldt from on tour, but neither Fredrika nor Mme. Armfeldt are happy about Desiree's career ("The Glamorous Life").

Scene Two. Anne and Fredrik enter the theater and sit in a box near the stage. Anne is now curious and suspicious about Desiree. The play begins, and Desiree makes a grand entrance. She sees Fredrik ("Remember?"). As the play continues, Anne insists that Desiree looked at their box and smiled. She starts to cry and runs off, followed by Fredrik.

Scene Three. In the Egerman house, Petra is straightening her blouse on the couch, Henrik is pulling on his trousers: They have sinned, and it was a complete failure. Anne enters, crying, and runs past them into the bedroom, followed by Fredrik. In the bedroom, Anne starts to reminisce about how when she was a little girl Fredrik would tell her stories until she fell asleep, and she called him "Uncle Fredrik." She speculates that Desiree must be very old. Fredrik says he will sit up for a while, and leaves the bedroom.

Scene Four. Desiree's rooms. Fredrik enters, having been told at the theater where to find Desiree. It has been fourteen years since they last saw each other. Desiree asks why he is there. He is not sure—curiosity? to boast? to complain?— and tells her of his erotic dream about her. He ventures that she must be lonely, and she tells him about her current lover, a married soldier. He notices a picture of Fredrika; Desiree tells him that Mme. Armfeldt insisted on having Fredrika live with her, rather than following Desiree around on tour. They are both hesitant to go on, so Fredrik starts to talk about Anne ("You Must Meet My Wife").

Although he is unable to say so, Desiree realizes that Fredrik wants to sleep with her, and takes him into her bedroom.

Mme. Armfeldt reminisces about some of the most profitable of her past relationships, in contrast to Desiree's sloppy affairs with men ("Liaisons").

In Desiree's room, Fredrik in a bathrobe and Desiree in a negligee come to the door, having heard her lover, Count Carl-Magnus Malcolm, swearing at his driver outside. She warns Fredrik that Carl-Magnus is very jealous and dangerous. Carl-Magnus enters; he has twenty hours' leave, of which nine are for Desiree and five for his wife. He wants to put on his robe, but it seems to be occupied. He is polite but icy with Fredrik, who quickly spins a story about being Mme. Armfeldt's lawyer and falling into a hip-bath. Desiree goes out to see if Fredrik's clothes are dry, and Carl-Magnus mentions his proficiency at dueling, which he demonstrates by throwing a fruit knife at a target. Desiree returns, having dipped Fredrik's clothes in her bath; Carl-Magnus offers him a nightshirt, but insists that he leave immediately. Carl-Magnus refuses to believe that Desiree would cheat on him ("In Praise of Women").

Scene Five. The breakfast room in the Malcolm house, where Charlotte is eating as Carl-Magnus enters. Her conversation is witty and bitter. She asks him how Desiree was. He tells her about Fredrik in the nightshirt, remembers that her sister is a school friend of Anne Egerman, and suggests that she pay Anne a visit, and tell Anne about her husband's activities.

Scene Six. In her bedroom, Anne talks to Petra, who is brushing her hair, about sex and men. She jokes about whether her body or Petra's is better-looking, and they are wrestling playfully as the doorbell rings. Petra answers the door: It is Charlotte Malcolm. They briefly catch up, then Charlotte turns to the subject of her husband. She begins to cry, cursing Desiree Armfeldt, who has seduced Carl-Magnus and now Fredrik as well. Charlotte tells Anne of the nightshirt incident, although she is ashamed to be carrying out her husband's wishes. Anne, realizing that Charlotte is telling the truth, begins to cry as well ("Every Day a Little Death"). Henrik enters and Charlotte leaves. Henrik asks what the trouble is, and Anne bursts into tears again. He tells her that it hurts him to see her unhappy, but her mood suddenly changes and she gaily pulls away and runs off.

Scene Seven. Mme. Armfeldt is playing solitaire on her terrace again, as Fredrika practices her scales. Desiree arrives. She asks her mother to invite the Egermans for the weekend, and reluctantly Mme. Armfeldt agrees. A montage follows ("A Weekend in the Country"): Anne is excited by the invitation until she realizes it involves Desiree; Fredrik is interested, but when Anne refuses to go, he agrees to decline for both of them; Anne tells Charlotte about it, but Charlotte persuades her that if she goes, she can win Fredrik back by making Desiree look old; Charlotte tells Carl-Magnus about the Egermans' invitation, and he promptly insists that they also go, uninvited; and Henrik decides to go to observe the world's snares. In the midst of all this, Desiree asks Fredrika how she would like to have a new father, the lawyer Egerman.

Act Two. The Entr'acte segues without a break into a song by the Quintet about the long days and short nights of the Scandinavian summer ("The Sun Won't Set").

Scene One. Mme. Armfeldt and family are idling on the lawn of her estate when they hear the horn of an automobile, well before the Egermans are due to arrive. They hurry into the house, and a moment later the Malcolms drive up. Carl-Magnus tells Charlotte to watch Fredrik carefully, then the Egermans arrive in their own car. The two couples greet each other cautiously. Petra unloads the luggage along with Frid. Desiree comes out to welcome the Egermans and is nonplussed to see the Malcolms, who give her a flimsy story about needing a place to stay for the night. The greetings over, Fredrika leads the guests into the house; both Fredrik and Carl-Magnus want to talk to Desiree privately, but she puts them both off.

Scene Two. Another part of the Armfeldt grounds, where Anne and Charlotte enter. Charlotte tells Anne she has a plan: She will flirt with Fredrik and make Carl-Magnus jealous enough to return to her. Fredrik enters and Charlotte immediately begins to play up to him.

Scene Two-A. Henrik tells Fredrika that he is hopelessly in love with Anne.

Scene Three. Fredrik and Carl-Magnus, both dressed formally for dinner, are pacing on the terrace ("It Would Have Been Wonderful"). Fredrika enters and tells Carl-Magnus that her mother wants to talk with him. As soon as they leave, Desiree enters. Fredrik begs her not to mention the nightshirt incident; he believes that Anne knows nothing about it. They hear Carl-Magnus calling for Desiree, and she invites Fredrik to talk to her later, in her room. He hides as Carl-Magnus enters and demands to know why the Egermans are there. Desiree continues the fiction about Fredrik being an old friend of her mother. Despite her protests, he promises to visit her bedroom as soon as possible that night. Frid announces dinner. Fredrik comes out of hiding, and Desiree takes him and Carl-Magnus by the arm and leads them in.

Scene Four. Mme. Armfeldt's dining room, where she sits opposite the guests ("Perpetual Anticipation"). The guests converse, Charlotte continuing to flatter Fredrik, to Carl-Magnus's displeasure. Charlotte starts to get drunk and Carl-Magnus orders her to leave the table. Henrik is disgusted and smashes his wineglass. When Fredrik reproves him, he angrily retorts that he despises his father and all of them. The others gently ridicule his earnestness, and he runs away from the table. Anne starts to follow him, but Fredrik orders her to come back.

Scene Five. In the garden, Henrik tells Fredrika that he is ashamed of himself. He hears Anne calling him and runs off. Anne comes on looking for him. She mocks his rigidity, but Fredrika tells her that Henrik is in love with her. Surprised, amused, and flattered, Anne goes off with Fredrika to find him.

Scene Five-A. Elsewhere on the grounds, Frid runs on followed by Petra. They kiss, but are temporarily interrupted as Fredrika and Anne cross the stage looking for Henrik. Then Frid and Petra start to make love. Henrik sees them and runs into the house.

Scene Six. In her bedroom, Desiree is sewing a rip in her skirt. Fredrik enters, having followed her. They laugh together about the situation. Then Desiree stops, disgusted with herself. She says that Henrik was right, they are both ridiculous. She tells Fredrik that she invited him because she thought that they could help each other make a real and "coherent" life together. Fredrik acknowledges the possibility, but admits that he is still infatuated with Anne. Desiree is crushed but gracious ("Send in the Clowns"). Fredrik apologizes for coming and leaves.

Scene Seven. The grounds. Henrik comes out of the house with a rope. He tries to hang himself from a tree branch, but the branch breaks and he falls. Anne finds him. They start to kiss each other. He tells her that he loves her, and she realizes that it is him she loves, not Fredrik. They start to make love.

Elsewhere, Frid and Petra have finished, and Frid is sleeping. Petra has fantasies about marrying a rich man one day, but in the meantime she wants to enjoy life while she can ("The Miller's Son").

Scene Eight. Elsewhere on the grounds. Charlotte is on a bench, crying. Fredrik enters looking for Anne. Charlotte apologizes for throwing herself at him, explaining that it was only to make her husband jealous. Then they see Anne and Henrik kissing and carrying their luggage to the stables to ride off together. Fredrik is devastated, Charlotte sympathetic.

Meanwhile, Carl-Magnus enters Desiree's bedroom and begins to undress, ignoring her demands that he leave. He ridicules the idea that Charlotte would ever cheat on him, much less with Fredrik. Then through the window he sees Charlotte on the bench with Fredrik. He rushes out.

On the bench, Charlotte apologizes to Fredrik for her meddling. Carl-Magnus, holding a pistol, comes out of the house and demands that Fredrik accompany him to the pavilion to play Russian roulette. Still dazed, Fredrik agrees. Charlotte is thrilled that Carl-Magnus is jealous.

There is a shot offstage. Desiree runs out of the house and Charlotte tells her what has happened. Desiree starts to run to the pavilion when Carl-Magnus enters carrying Fredrik, whom he tosses to the ground. Desiree is horrified, but he says that Fredrik missed and merely grazed himself. He tells Charlotte that he will forgive her, but they must leave immediately. She agrees and they kiss and leave. Desiree awakens Fredrik. He tells her about Henrik and Anne, and is astonished to discover that he feels rather relieved. He jumps up and tells Desiree that he wants the coherent life she suggested, with her and Fredrika. They kiss.

Elsewhere, Fredrika tells her grandmother that she hasn't seen the night smiling, but Mme. Armfeldt retorts that the night has already smiled twice, for the young and for the fools, and only the last smile is to come. She dies. All of the other main characters, now with their proper partners, waltz on again, until the Liebeslieder singer who started the show concludes it by playing a chord on the piano.

It is clear from these summaries that the libretto of the musical has followed the original movie in numerous ways, large and small. Although according to its creators *A Little Night Music* underwent several changes of tone and style, and many early songs were replaced, the basic plot changed very little, and many details and lines of dialogue were kept or changed only slightly. The humor of the dialogue in the musical tends to be wittier, and a little more obvious. Besides reflecting differences of taste, this probably also reflects the adapters' need to keep the audience interested and amused throughout a live performance.

Virtually all of the speaking parts remain the same, with one interesting exception. This is, of course, the change in age and gender of Desiree's child, from four-year old Fredrik to thirteen-year-old Fredrika. Although it may seem like an unnecessary alteration, it accomplishes a small but important goal in the musical. It adds another speaking (and singing) character, to whom other characters reveal things, and who herself becomes pivotal in a small way when in Act Two, Scene Five she tells Anne that Henrik loves her. This scene, replacing the mechanical moving bed of the film, saves the musical a good deal of time, money, and contrivance.

In many small ways the story has been improved in terms of logic and momentum. For instance, in the movie it is strange that Count Malcolm, who has just arrived home, knows about the invitation to Mrs. Armfeldt's estate not only before his wife, but also long before the invitations could possibly have been sent. In other places, logic may have suffered in the adaptation. For instance, in Act Two, Scene Eight, Carl-Magnus is confident that Charlotte would never cheat on him, but suddenly he is convinced she is cheating simply because he sees her on a bench with Fredrik. Those familiar with Bergman's movie may also notice that many of its nuances and ambiguities have been replaced, perhaps necessarily, by more obvious statements and devices, and that a number of titillating irrelevancies have been added, such as the hint of lesbian attraction in Anne and Petra's wrestling in Act One, Scene Six.

A number of the musical's songs originated in the dialogue or story of the movie. Other songs—such as "The Sun Won't Set," which is used almost as a leitmotif throughout the show, and "Perpetual Anticipation"—seem to have been inspired by its moody lighting, camerawork, and black-and-white photography. These aspects of film are unique to the medium, and many of the nuances of Bergman's direction (such as the subtle emphasis on the moon shining through the castle windows and the strange shadows it casts), if they were not to be dropped completely, had to be translated into another language of mood and atmosphere. Sondheim has accomplished this in his score.

ADAPTATIONS FROM OTHER LITERARY GENRES

A nondramatic work is usually much more difficult to adapt than a dramatic one. In literary fiction, for instance, much of what the main characters experience is internal—psychological and emotional—which makes it extremely difficult to translate into theatrical terms. Some internal monologues, of course, can be translated into soliloquy songs, but to have more than a couple of these in a show would create monotony. Therefore literary sources need more drastic changes than dramatic ones. When a nondramatic work is being transformed into a dramatic work, the chief problem is to ensure that the result contains the essentials of drama: action, conflict, characters who change, variety, the dramatic arc or some other workable structure, and so on.

Even more problematic as a basis for a musical than fiction, which usually provides at least a plot, is nonfiction such as biography or history. It might seem that biographies offer natural fodder for musical adaptation, since a person's life story would provide both a protagonist and a plot outline. But upon consideration it becomes clear that biographies are perhaps the most intractable sources of material a writer can choose. First, biographies are usually about famous people. If a man is or was famous, many people in the audience will know details of his life. This restricts the amount of shaping and transformation that writers can do to make him sympathetic, or to make the plot clearly structured and effective. A second and related problem is that most people's life stories rarely lay out in a way that resembles the dramatic arc or other convenient plot structures. Finally, in the case of a man who is alive or recently deceased, he or his estate may put additional restrictions upon what aspects of his life can be changed, or for that matter, what can be shown and discussed onstage. The record books are filled with biographical musicals that either flopped quickly, or succeeded commercially but not artistically. In fact, there are only three biographical musicals that have succeeded in both ways: *Annie Get Your Gun, Gypsy*, and *Fiorello!* But *Annie Get Your Gun* retains only a few facts about Annie Oakley's life. It changes almost everything else—for instance, that her real name was Phoebe Moses and that she married Frank Butler long before either one joined the Wild West Show. *Gypsy* is not based upon, but "suggested" by the memoirs of Gypsy Rose Lee, because librettist Arthur Laurents came to realize that she had made up much of the material in her book. As a result, he felt free to invent much of the story, and indeed he even invented the leading male role, Herbie. As for *Fiorello!*, despite the high quality of its writing and production, it has not held up well over time; it now seems like a well-made but old-fashioned and formulaic show, due to the limitations of trying to fit a real man's life into the conventions of a musical comedy.

Writers face a still greater degree of difficulty when adapting history, rather than biography. When a historical book or event is used for a musical, the writers must choose which of the many people involved to use as characters, and which of those will be the protagonists. In addition, few historical accounts have an inherent dramatic structure, and therefore a structure needs to be superimposed. Few Broadway musicals have been faithfully based upon historical literature or events, and of these only *1776* has been successful.

Other types of nonfiction hold even more challenges for adaptation than do biography or history. But at least one first-class musical resulted from such an adaptation: *How to Succeed in Business without Really Trying*. The original source by Shepherd Mead was a satirical self-help book, offering advice to unscrupulous corporate climbers on becoming rich and powerful. The adapters took the name "Pierrepont Finch," used in all of Mead's sample situations as a name for such a climber, and created a protagonist named J. Pierrepont Finch, who reads a book called *How to Succeed* and follows its advice. Structuring the story around his rise to the top of the corporate ladder, they used a number of situations from the book as the bases of scenes and songs such as "Grand Old Ivy." They also took from Mead's book the name and rank of the head of the company, J. B. Biggley; other names such as Mr. Bratt, Mr. Gatch, and Hedy; and phrases such as "wicket," "Chipmunks," and "A Secretary Is Not a Toy."

Finally, we should mention poetry as a basis for adaptation. The most likely poetic source for a musical is an epic or narrative poem, which like a novel provides a story and a set of characters. For instance, *The Iliad* and *The Odyssey*, radically transformed and set in Washington state at the turn of the twentieth century, were the sources for *The Golden Apple*. Nevertheless, there have been adaptations of other kinds of poetry: *Shinbone Alley* was based on the epistles and free-verse poetry of Don Marquis's Archy and Mehitabel; *The Wild Party*, of which there were two unsuccessful adaptations in the same year, was based on the poem by Joseph Moncure March; and of course *Cats* was based on T. S. Eliot's *Old Possum's Book of Practical Cats*.

As an example of the adaptation of a literary work we will look at James Michener's *Tales of the South Pacific*, adapted by director and librettist Joshua Logan, lyricist and librettist Oscar Hammerstein II, and composer Richard Rodgers into the musical *South Pacific*. Michener's book contains a group of interconnected stories about American sailors and marines serving in the New Hebrides, Solomon, and Santa Cruz islands between 1942 and 1944, in the middle of World War II. As before, we will compare synopses of the book and the musical. (Discrepancies in spelling and nomenclature in these synopses follow the discrepancies between the book and the musical.)

Tales of the South Pacific

The stories in this book present a tapestry of life on the American-held islands of the South Pacific during World War II. Many characters appear in more than one story; the main character of one story often plays a small part in another. Logan and Hammerstein chose two stories in particular, "Our Heroine" and "Fo' Dolla," for the core of their libretto. "Our Heroine" is about Nellie Forbush, a pretty young nurse from Arkansas. Another story, "An Officer and a Gentleman," tells how she had previously fallen in love with a handsome, snobbish lieutenant from a privileged background, dating him until he coldly told her that he was married. In "Our Heroine" she is sent north to a new hospital. On her first night she is invited to a dinner at a local French plantation in honor of the new nurses, where she meets Emile De Becque, a handsome older plantation owner with a mysterious past. She is immediately interested in him and his story, and sees him at other dinners over the next few weeks. Finally De Becque invites some of the Americans, including Nellie, to his own plantation for dinner, and as they are leaving he invites Nellie to visit again, promising to show her his cacao grove. She comes back several days later, and they sit in a pavilion in the grove, talking idly. He asks her to marry him, and although she has already decided to, she takes a few days to think it over before giving him an answer.

De Becque goes off for a few days to deliver supplies to Bali-ha'i, a nearby island. While he is gone Nellie goes to a dinner where a guest talks about De Becque's daughter, a half-Javanese young woman with a colorful life. De Becque has three other daughters by Javanese women, and four younger daughters whose mothers were Polynesian and Tonkinese (Vietnamese)—all of them illegitimate, as De Becque himself tells Nellie when he returns. She tries to take this news in stride, but the thought that De Becque has lived with and had children with a Polynesian, whom she considers a "nigger," is revolting to her, and she turns down his marriage proposal. That night, talking to an older nurse, she recalls that she left

Arkansas and came to the South Pacific to learn new things and meet new people, and decides that she will marry De Becque after all. She takes a jeep out to De Becque's plantation and finds him singing a folk song with his small daughters. She joins in with them.

The longest story in the book, "Fo' Dolla," is about Joe Cable, a young Marine lieutenant from upper-class Philadelphia, who is assigned to stop Bloody Mary, a colorful, foulmouthed old Tonkinese woman, from selling grass skirts to American servicemen. Mary has defied all previous attempts to stop her operation, but she likes Cable, and he finds her interesting. Eventually she takes him with her to Bali-ha'i, the mysterious island where her family lives, and where the French have sequestered all the unmarried French and native women. There she leads him to a house where he meets Liat, her young and beautiful daughter. Mary leaves them, and they make love. Over the next few days Cable is unable to talk to anyone about what has happened, or to write to his girlfriend at Bryn Mawr. He makes two more trips to Bali-ha'i, and each time he sleeps with Liat. On his second trip, the nun who runs the local hospital tells him that he should stop seeing Liat, who is wanted in marriage by a wealthy French planter, and that if Cable returns, she will report him to his commander for his own good.

Back at the base, Cable finds his work deteriorating. When Bloody Mary confronts him, he tells her that he loves Liat but can't marry her. Mary gets angry and he slaps her. She offers him all her money if he will marry Liat and live there. He runs away from her, but over the next few weeks he continues secretly to see Liat, who knows that he will not marry her but does not care. Then Cable gets new orders: A big offensive is on, and all forces will be moving north. He goes back to Bali-ha'i to see Liat one last time, but she has left to marry the planter. Bloody Mary takes him to Liat and they make love one last time. He gives her his watch, but they hear the planter's car approaching, and he runs off. The next day he leaves with his unit. Bloody Mary smashes his watch on the road and goes back to selling her wares to servicemen.

There are other stories involving these characters that the adapters omitted, and some that they incorporated in part, such as "Dry Rot," the story of a sailor who, while stuck with maintenance duties on a tiny atoll for two years, meets and is befriended by Luther Billis, a big, gregarious, tattooed, wheeling-dealing Navy construction worker. "A Boar's Tooth" is about Billis and how he finagles a trip with some officers to Vanicoro, a mysterious island with twin volcanoes near Bali-ha'i, where they witness a native ceremony involving the tusks of wild boars. "Alligator" is about Operation Alligator, the big counterstrike planned by the Pentagon to drive the Japanese out of the South Seas. "The Cave" is about a man who risks his life every day behind enemy lines watching the coasts and radioing news of Japanese movements to the Allies, until he is tracked down and killed by the Japanese. And the final story, "A Cemetery at Hoga Point," reveals that Joe Cable has been killed in action.

SOUTH PACIFIC

Act One. Scene One. A pagoda and terrace on Emile de Becque's plantation. A young Eurasian boy and girl are playing and singing ("Dites-Moi"). A servant leads them off. De Becque enters from lunch, showing Nellie Forbush his estate. They have known each other for two weeks, and they talk about themselves ("A Cockeyed Optimist") as they are falling in love ("Twin Soliloquies"). He asks her to marry him ("Some Enchanted Evening"), and although she is clearly inclined to, she asks for a few days to think it over before giving him an answer. When she leaves, the children return and we learn that de Becque is their father.

Scene Two (in one). We meet Bloody Mary amid a group of sailors and Marines ("Bloody Mary Is the Girl I Love").

Scene Three. Immediately afterward, on a palm grove near the beach. We see Mary's kiosk and Luther Billis's laundry operation in action. Billis trades the grass skirts that he and his men have made to Mary in exchange for a boar's tooth bracelet. He would love to go to Bali Ha'i, a mysterious island with two volcanoes, where he knows he could get boar's teeth and other valuable objects, but he needs an officer to get a boat. The nurses jog by, and Luther gives Nellie her laundry ("There Is Nothing Like a Dame"). Then Joe Cable enters, looking for Emile de Becque. Bloody Mary, who finds Cable "damn saxy," tries to sell him some of her trinkets. She tells him about the special magic of Bali Ha'i ("Bali Ha'i"), and Billis tries to convince him to take a boat there. Enter Captain Brackett and his executive officer Commander Harbison, who tell Mary to close her operation. She refuses, but when Cable starts to pack it up, she lets him. Cable reports to Captain Brackett, explaining that he is there to establish a coast watch behind enemy lines. He needs a local to help him, and has been told that de Becque knows the islands well.

Scene Four (in one). Billis has arranged a boat for Bali Ha'i but needs Cable to go with him. Cable refuses.

Scene Five. Brackett's office. Brackett and Harbison have invited Nellie there, hoping that she can get information about de Becque for them. If de Becque is trustworthy, he and Cable might be able to get information the Allies need to make Operation Alligator a success. Nellie realizes that she doesn't know very much about de Becque.

Scene Six (in one). Nellie and Cable meet and discuss the letters of advice they get from their mothers.

Scene Seven. The beach, where Billis has built a pay shower stall. Nellie starts to shower and tells her fellow nurses that she has decided not to marry de Becque, because she knows so little about him ("I'm Gonna Wash That Man Right Outa My Hair"). After her shower, though, de Becque enters. He has planned a party to introduce Nellie to his friends, where she can learn more about him. She asks him about his past, and when he answers her questions frankly, she agrees to go to his party ("I'm in Love with a Wonderful Guy").

Scene Eight. Brackett's office. Brackett and Harbison have asked de Becque to join Cable on his mission. De Becque refuses, because it is too risky, and the life he plans with Nellie is more important to him than anything else. He and Brackett leave, and Harbison suggests to Cable that he take a few days off to await further developments. Cable decides to accept Billis's offer of a boat.

Scene Nine (in one). Bali Ha'i, where Billis, Cable, and Bloody Mary have come ashore. Mary arranges for Billis to be taken to the boar's tooth ceremony, and leads Cable away.

Scene Ten. The interior of Mary's hut. Mary leads Cable in. Liat enters and Mary leaves. Cable and Liat, who are instantly attracted to each other, embrace and he starts to undress her. Later, after they have made love, she tells him that Bloody Mary is her mother. He hears the boat bell ring and reluctantly leaves ("Younger than Springtime").

Scene Eleven (in one). Cable joins Billis and Mary on the shore. He and Billis leave, and Mary tells the other natives that he will be her son-in-law.

Scene Twelve. De Becque's terrace. The guests are leaving his party. He and Nellie are very happy. The two children enter, and he tells Nellie that they are his. Shocked, Nellie tells him that she has to get back to base, and despite his entreaties, she leaves hurriedly.

Act Two. Scene One. A Thanksgiving stage show on the base. A big group dance number ends, and Nellie is mistress of ceremonies.

Scene Two. Backstage, immediately afterward, where Billis is stage-managing. De Becque enters, looking for Nellie; Billis tells him that she has asked for a transfer. Cable enters. He has been sick for several days, and unable to get to Bali Ha'i. Mary and Liat enter, and Mary tells Cable that a wealthy French planter wants to marry Liat. She offers him all her money if he will marry Liat and live there ("Happy Talk"). He gives Liat his watch, but says that he can't marry her. Mary smashes his watch on the ground and takes Liat away to marry the planter.

Scene Three. The stage, immediately afterward. Captain Brackett thanks the performers. Commander Harbison announces the finale, a drag number performed by Nellie, Billis, and the chorus line of nurses ("Honey Bun").

Scene Four. Backstage, immediately afterward. Nellie and Cable talk briefly. De Becque enters and asks Nellie if they can talk. She asks Cable to stay. She tells de

Becque that she can't marry him, she can't help the way she feels. She runs off. Cable asks de Becque to reconsider joining him on his mission ("You've Got to Be Carefully Taught"), and de Becque agrees ("This Nearly Was Mine").

Scene Five (in one). A naval airplane has just taken off.

Scene Six. The base radio shack. Brackett is anxiously waiting for a signal from Cable and de Becque. Harbison enters with Billis. Suddenly they hear de Becque's voice on the radio. He and Cable begin to radio news of Japanese movements, to Brackett's delight.

Scene Seven (in one). A montage of two short scenes, as de Becque and Cable continue to radio information over the next two weeks to the Allies.

Scene Eight. The radio shack, shortly afterward. Nellie asks Brackett for news of de Becque, whom she has been trying to find. Brackett tells her about the mission, and they hear de Becque's voice on the radio, telling them that Cable has been killed. He adds that the Japanese are preparing to pull out, but are searching for him in airplanes. The sound of a plane is heard and de Becque signs off abruptly. Brackett assures Nellie that he may survive.

Scene Nine (in one). A dance is being held, but Nellie walks past it.

Scene Ten. The beach, immediately afterward. Nellie now realizes that the only thing that matters is being with de Becque. Bloody Mary enters with Liat and asks Nellie where Cable is, saying that Liat refuses to marry anyone else. Mary embraces Liat consolingly.

Scene Eleven (in one). The base is preparing to move, but Billis is still trying to make a buck. When Brackett enters, Billis volunteers to join a mission to rescue de Becque. Brackett tells him that Operation Alligator has already begun, and a plane has been sent to try to find de Becque.

Scene Twelve. De Becque's terrace. Nellie is there with his two children, watching the ships and airplanes leave. Nellie sits them down to eat as de Becque enters unseen. Nellie and the children start to sing and de Becque joins them. Nellie serves him food and they clasp hands.

While much of the work of the adaptation is easy to see in these summaries, there are a few particulars worth mentioning. Many of the essentials of *South Pacific* are straight out of the book: the milieu, the main characters and their stories, the background of Operation Alligator, and the twin themes of racism and of love bridging racial and cultural divides. Probably the most important choice made by the librettists was to recognize the similarity of the themes of "Our Heroine" (the Nellie-Emile story) and "Fo' Dolla" (the Cable-Liat story), and to emphasize the parallels in their adaptation. They did this by taking the former story for the main plot and the latter for the subplot, and condensing both stories while preserving the essence of each. They also incorporated the central idea of "The Cave" as a crucial pivot to tie together the two plots—which had the additional benefit of giving the rather flimsy story of "Our Heroine" additional suspense and plot twists before its resolution. In addition, they wisely combined Vanicoro, the mysterious island of the boar's tooth ceremony, with nearby Bali-ha'i, the island of Bloody Mary and Liat, and enlarged the significance of Luther Billis as comic relief and a male foil for Bloody Mary. In the process, while creating a large number of new characters, they also retained several characters from the stories in small roles (such as Dinah, a nurse who befriends Nellie, and Bus Adams, an aviator), and condensed de Becque's eight daughters into one son and one daughter. Captain Brackett and Commander Harbison are convenient stand-ins for all the commanding officers who appear or are mentioned in the stories; the adapters took Harbison's name from the main character of "An Officer and a Gentleman," the lieutenant who dumped Nellie before she met De Becque.

Again, many ideas for songs in the musical can be found in the original stories. For instance, early in the cacao-grove scene of "Our Heroine" there is a moment when Nellie is thinking about marrying Emile and living there, and Emile is trying to gather the courage to ask her—a moment that Rodgers and Hammerstein turned into the memorable "Twin Soliloquies." Other songs, such as "Happy Talk" and "There Is Nothing Like a Dame," encapsulate a mood or emotion that is implied in the original.

ADAPTATIONS OF MYTH AND FOLKLORE

While myths, legends, folk tales, and fairy tales are often considered branches of literature, many such stories have been so widely disseminated that they cannot be considered to exist in only one literary version. But the problems for adaptation posed by myths or legends are similar to those of other types of literature, compounded somewhat by the potential difficulties of representing supernatural beings and events onstage. Another problem with this material is that in many myths and folktales the focus is not on characterization but on incident, and many of the characters do not have the depth of the characters in the best drama and fiction. Therefore, adapters need to develop or create the nuances necessary to give their mythical characters recognizable personalities.

Despite this, there have been quite a few musicals based on mythological and legendary material. Examples include *Out of This World, The Golden Apple, Cinderella,* and *Once Upon a Mattress. Beauty and the Beast* and *The Lion King* are also essentially adaptations of folklore—indirectly, since they both started as Disney movies. *One Touch of Venus* and *Into the Woods* could also be considered to fall into this category, although in both cases the original legends provide only a small part of the story lines.

ORIGINALS

Any musical not based on an existing dramatic or literary work is called an "original," although the term is slightly misleading. No idea is entirely original, and no work of art is ever created in a vacuum. Because of the many difficulties involved in imposing the strictures of musical theater on new ideas, comparatively few original musicals have become successful—but among that number are *Brigadoon, Finian's Rainbow, Bye Bye Birdie, Company, A Chorus Line,* and *Sunday in the Park with George.*

It is in the creation of an original show that knowledge of, and experience in, dramatic writing are most crucial. There are no guidelines and no rules for creating an original, except the ones already discussed. All of the dramatic principles mentioned earlier—such as the expression of a conflict through specific human behavior, a structure utilizing the dramatic arc or a similar shape, the early introduction of likeable leading characters who grow or change, variety, comedy, and showing rather than telling—must be incorporated into the material. Although we can't compare the libretto of an original musical with written source material as we did for *A Little Night Music* and *South Pacific,* we can point out some salient features of two originals that became hit shows.

Alan Jay Lerner always called *Brigadoon* an original musical, and claimed that it originated in his conviction that faith could move mountains. Almost from the beginning, however, people have claimed that *Brigadoon* was based on a short story named "Germelshausen," by the nineteenth-century German-American writer Friedrich Gerstäcker. There are indeed some striking resemblances between its plot and that of *Brigadoon*: Like the musical, "Germelshausen" is about a little village that magically appears once every hundred years, and an outsider protagonist who meets and falls in love with a village girl. Aside from that, however, the two stories have nothing in common. In "Germelshausen," the village's magic is not the result of a blessed miracle, but of evil and a curse, and people only stay if they are inveigled to do so under false pretenses. Nor does the protagonist learn the truth until he has left the village, never to return. Equally important, the story has none of the characters or subplots of the musical, and the milieus are very different: Germelshausen is a medieval German village, Brigadoon a seventeenth-century Scottish one. Finally, the story has no trace of the theme of romance versus cynicism that is the core of the musical. Even if Lerner did read "Germelshausen," almost everything in the libretto of *Brigadoon* is his own.

A Chorus Line is another celebrated original—but, as many people know, it began with two all-night sessions of interviews and reminiscences by a group of Broadway dancers. Some of the participants in these sessions became original cast members, and many of the stories, lines of dialogue, and lyrics of the musical came from them.

Decades after its premiere, it is easy to forget how audacious the libretto of *A Chorus Line* was. There is no single protagonist, although Cassie and Zach are featured somewhat more prominently than the other characters. Virtually the entire show is performed presentationally (toward the audience), since Zach spends most of the show hidden behind the audience, and interaction among the other characters is kept to a minimum. In addition, the plot is extremely simple and there is no subplot. Much more unusual, however, is the fact that there is no real story. The dancers are there to audition, and they do, but the greater part of the show consists of them talking about themselves and recounting moments from their lives. There are not even any true flashback scenes; most of the recalled material is presented narratively. Given its origins it would have been easy for *A Chorus Line*, like many of its unsuccessful imitators in the 1970s and 1980s, to have turned into a revue, and in this light its achievement as a book musical becomes even more impressive. The creators made many smart choices. The overall concept of an audition for a Broadway musical, as obvious as it now seems, was not their first idea, but it proved an effective framework for the show as a whole. They also made use of many different modes of storytelling, to avoid the monotony into which the telling of similar stories could easily have fallen. In addition, they cleverly arranged the various dancers' memories in a chronological sequence that leads from early childhood (in their earliest memories and "I Can Do That") through adolescence (in the "Montage"), to their professional life (in "Dance: Ten, Looks: Three" and Paul's monologue), their present situations (in "The Music and the Mirror" and "The Tap Combination"), and their potential futures (in "What I Did for Love"). Especially impressive as a narrative and structural device is the immense central "Montage," which—though centered around the song "Hello Twelve, Hello Thirteen" and the agonies of adolescence—daringly leaps from one character's memories to another's, often blending several together simultaneously, in a dazzlingly unpredictable melange of almost all the modes of storytelling to be found in musical theater: movement, dance, spoken monologue, recitative, song fragments, mini-songs (Maggie's "Mother"), and even a full-length song (Diana's "Nothing").

However, none of these daring innovations might have been made to work without a uniquely long time to develop them. *A Chorus Line* had almost six months of workshops, rehearsals, and previews in which to evolve. Most

Broadway shows have five weeks, and many smaller productions have less time than that. Of course, original shows do not necessarily require six months of rehearsals. But they usually need considerably more preparation than adaptations, and the more original the project, the more time it requires.

SUBPLOT

Subplot is, as its name indicates, a plot that is both different from and subordinate to the main plot of a dramatic work. Subplot was long a staple of musical theater, deriving from its ancestry in opera and operetta; it can also be found in many American operettas such as *Rose-Marie* and *The Desert Song*, and in musicals well into the 1940s and even the 1950s. Such shows as *Oklahoma!, Carousel, Brigadoon,* and *The Most Happy Fella* adhere to this convention or a close variant of it.

Why were subplots once so popular? Several reasons come to mind. In Viennese operetta, the need for variety was well understood. The two romantic leads, as a rule, were good-looking and sang beautifully, but they were usually not funny, so another subsidiary couple was used to handle the comedic elements of the story. This provided variety in two ways: a second couple to complement the lead couple, and comedy to complement the romance. It may have also been felt that the main plots of most operettas were not interesting enough to hold the stage for a full evening, and so a parallel or complementary subplot was necessary. Early twentieth-century musicals, which evolved in part from operettas, often had a similar division of labor between the singers and the comedians.

In good musicals, the subplot is not simply another group of characters. It is another plot with its own arc and resolution, although it is related in some way to the main plot. The best subplots are tightly integrated with and complement the main plot, sometimes thematically, and sometimes by supplying what the main plot lacks—humor, a different age group, dancing rather than singing, a sad resolution rather than a happy one, and so on—and each plot parallels or works in counterpoint to the other. Good examples of this include the Bernardo-Anita subplot in *West Side Story* and the Herr Schultz-Fräulein Schneider subplot in *Cabaret.*

By the 1940s there was considerable variety in the implementation of subplots. For example, in *Brigadoon* the Charlie-Jeannie-Harry triangle is the subplot. But perhaps because, like the main plot, this subplot was a serious and lyrical one, the comedic relationship between Jeff and Meg Brockie could also be said to form the makings of a subplot. The same is true of *South Pacific:* The subplot is the Cable-Liat relationship, but again, this subplot is lyrical and serious, and the bantering between Bloody Mary and Luther Billis could be considered the torso of another, comedic subplot.

After the 1950s, writers became even more creative in their use of subplot. The leading pair could be funny as well as romantic, and so could the subsidiary pair; it was merely a question of finding performers who could do both. The rising cost of musicals may have also increased the necessity of finding performers who were good at both comedy and romance. In *Guys and Dolls*, the two couples and plots (Nathan-Adelaide and Sky-Sarah) are equally important; in *Kiss Me, Kate*, both the lead couple (Fred and Lilli) and the secondary (Bill and Lois) are comedic, though only the lead couple is also romantic. In *My Fair Lady*, Alfred P. Doolittle and his cronies provide only the barest bones of a subplot, and a show like *How to Succeed in Business without Really Trying* from 1961 has virtually no subplot at all. When subplot was used in the 1960s and afterward, it tended to carry most of the weight of romance in the show; in *Hello, Dolly!* and *Fiddler on the Roof* the protagonists are middle-aged and their relationships are largely comedic, while the subplot of young romance is carried by the subsidiary couples. Shows of the 1970s like *Company* and *A Chorus Line* have no subplots. (For that matter, they barely have plots.) Nowadays, subplot seems to be unnecessary—but only when the main plot and the leading characters are interesting and compelling enough. Subplot can still be a useful structural device.

FUNCTIONS OF THEATER SONGS

The primary difference between songs written for a musical and "pop" songs written for listening or dancing is that theater songs serve specific *dramatic* purposes. Good theater songs move the show and its characters in their journeys, whether these journeys are physical, emotional, or intellectual. They can do this by setting the tone of the show, setting particular scenes, driving the plot forward, or revealing characters' backgrounds, principles, or desires. Only "list" songs do not show a journey, but even they can be effective in comic or light-hearted moments. (We'll discuss list songs, and other types of songs, in chapter 5.)

Experienced theater writers agree that the first ten minutes of any musical are the most crucial, because these first minutes tell the audience what to expect from the rest of the evening. In these ten minutes the opening scene must be set clearly, and so must the tone and style of the show as a whole. These tasks are often accomplished by an opening number.

Before 1943, the almost universal convention was to start a show with a big chorus number, which might or might not have any other function. Most shows began with a big, loud "showbiz" opening, no matter what the plot or style of the show might be. Even *Show Boat*, while ahead of its time in many ways, started with a big chorus number.

In 1943 *Oklahoma!* surprised Broadway veterans by opening with a solo ballad, "Oh, What a Beautiful Mornin'," that begins while the singer is offstage. *Oklahoma!* demonstrated that a show does not have to begin with a big "razzmatazz" number. For many shows, such an opening would be misleading and incongruous. Since *Oklahoma!*, writers have found increasingly novel and innovative ways to begin shows. *My Fair Lady* has no opening song, although it does have an opening number with movement, dancing, and juggling, and *Phantom of the Opera* begins without any music at all; even the brief overture comes after the opening auction scene.

Many musicals have failed because the opening promised something that the rest of the show did not deliver. Perhaps the most celebrated example of the power of an opening number to help or hurt a show is the case of *A Funny Thing Happened on the Way to the Forum*. Before it came to Broadway, it was previewing in Washington, D.C., to small, unenthusiastic audiences, when Jerome Robbins was brought in as a play doctor. He watched the show and told the staff that nothing significant needed to be changed except the opening number, a charming little soft-shoe by Stephen Sondheim called "Love Is in the Air," which Robbins said was all wrong for the show. *Forum*, of course, is in no way charming or little; it is a wild farce with elements of vaudeville and burlesque. Sondheim replaced "Love Is in the Air" with a big brassy number called "Comedy Tonight." This song told the audience exactly what kind of an evening it was about to see, and *Forum* became a hit.

In addition to the basic functions of the opening that we've already mentioned—setting the first scene and setting the tone and style of the show as a whole—there are two other tasks that many opening numbers accomplish. Often the opening introduces the leading characters and their situation, although this is not a necessity as long as they are introduced soon afterward. Examples of this introductory function are the "Christopher Street" opening of *Wonderful Town* and the "Comedy Tonight" opening of *A Funny Thing Happened on the Way to the Forum*. The other task is to present the theme or spine of the show. The classic example of an opening number that presents the theme, in addition to all the other functions noted, is the opening number of *Fiddler on the Roof*. We have already recounted how Jerome Robbins continued to demand that the writers verbalize the theme of the show, until one of them finally suggested that it was about the dissolution of a way of life. As with *Forum*, Robbins insisted that the opening number needed to prepare the audience, to show the traditions that were going to change. He wanted this opening to function like a tapestry against which the entire show would play. The result was "Tradition," which sets the scene and the tone, introduces the major characters, and presents the theme and spine of the entire show as well.

After the opening number, subsequent establishing numbers can occur at any major scene change, to set the new scene or mood. In Act One, Scene Seven of *My Fair Lady*, "The Ascot Gavotte" helps to set the new milieu of the club tent at a racetrack reserved for the upper class. At the beginning of the second scene of *South Pacific*, "Bloody Mary Is the Girl I Love" helps to change both the mood and the scene quickly from the quiet intimacy of de Becque's house to the noisy beach full of rowdy sailors and marines.

Songs can also advance the plot. (Of course, every song should further the story in *some* way. As Lehman Engel used to say, good writers don't want a song that "stops the show"—they want a song that moves it forward.) Sometimes this is done by setting all or part of a scene to music. In "number operas" it is common for the last ten or fifteen minutes of an act to be entirely set to music as a finale, with one tune or section leading into another without a pause. Typical examples are the finales of Act Two and Act Four from Mozart's *The Marriage of Figaro*. Gilbert and Sullivan's comic operas all follow this convention, as did many early musicals and operettas, such as *Of Thee I Sing*, which used Gilbert and Sullivan's operas as models.

In more recent musicals it has been common for an entire scene to be constructed around a single song. The long opening number of *A Chorus Line* is built around the song "I Hope I Get It," with dance interludes; and in Act Two, Scene Two of *Company*, the entire postcoital scene between Bobby and April consists of the duet "Barcelona."

An occasional practice is to summarize an offstage scene in a song. We have already mentioned the offstage scene in *My Fair Lady* that is recounted by Higgins and Pickering in "You Did It." Another effective example is "The Baseball Game" from Act Two of *You're a Good Man, Charlie Brown*. This song summarizes the big game by cutting back and forth between short vignettes of the game and Charlie Brown's subsequent description of it in a letter to a pen pal, all set to music. However, this technique is most effective when used sparingly. Since it is in the nature of a musical to have almost all the important events take place in front of the audience, writers ordinarily put an important scene offstage only if it is impractical to put it onstage, or if—as in the two examples here—the resulting song is more theatrical, interesting, or funny.

Yet another way to use song to advance the story is to take advantage of the presentational or nonlinear techniques that are possible in musical theater, such as *montage*, the combination or superimposition of short scenes to create the effect of bridging space or time. Montage is ubiquitous in film. It is hard to do in "straight" theater, but relatively easy in musicals, because of the nonrealistic and almost magical power of adding music. Just as music can help to bridge the gaps between scenes, it can also help to bridge gaps of time; a piece of music can provide a continuity that unites disparate scenes, making them appear as parts of a larger whole. We have mentioned the "Poor Professor Higgins" montage in Act One of *My Fair Lady* in which Higgins tries to teach Eliza to enunciate properly, and the enormous "Montage" number in *A Chorus Line* that intermingles the entire cast's memories of adolescence. Another classic sequence is the striptease montage at the end of Act Two, Scene Five of *Gypsy*. In only a few minutes we see the tremendous changes in Gypsy Rose Lee from her very first, tentative performance as a gawky unknown stripper in Wichita to her triumphant appearance in New York as a star, months or years later, all while performing "Let Me Entertain You." Many of Stephen Sondheim's later shows have made imaginative use of montage techniques. In *Follies*, "Waiting Around for the Girls Upstairs" leads the four principals into a flashback of thirty years ago, then at the end returns them to the present; in *A Little Night Music*, "Soon" weaves together the thoughts of the three Egermans in ironic counterpoint; and several songs in *Pacific Overtures* use montage to compress many events and years into minutes, as in "Please Hello" and "A Bowler Hat."

Songs can also help to move a show forward by revealing character more quickly and tellingly than is possible in a play. Such "I am" numbers acquaint us with a character, not only through the lyrics but also through the musical setting. When Linda Low sings "I Enjoy Being a Girl" in *Flower Drum Song*, we get a clear sense of her youthful enthusiasm and sassiness, not only from the lyrics but from the rhythms of the melody and the bouncy accompaniment. When Gittel Mosca sings "Nobody Does It Like Me" in *Seesaw* we get a picture of an older and more experienced woman. In both cases, the lyric reveals an essential aspect of a character, giving information that in a play would take an entire scene or an extended monologue; and in a good song, the music also tells us something about the character, like the flavor of her personality and her individual style. (We will discuss specific ways in which lyrics and music can reveal character in chapters 5 and 6.)

Although songs can reveal character by direct utterance, they can also reveal it indirectly. In *The King and I*, when Lady Thiang sings "Something Wonderful," she is telling Mrs. Anna and the audience about her husband's personality; but in doing so, she also reveals a great deal about herself—about her character and her feelings for her husband.

A song can also reveal a character's desires, or the motivation for an action about to be taken. "I want" songs, which usually come early in a show, propel the story forward more forcefully than "I am" songs. In addition, when characters tell us what they want, they also tell us what kind of people they are. Examples include Rose's first song in *Gypsy*, "Some People," which reveals her forceful and domineering personality; and "Skid Row," Seymour and Audrey's first song in *Little Shop of Horrors*. When they separately express similar "wants" in this song, it foreshadows their future relationship.

Songs can also serve as relief, either comic or otherwise. A perfect example is the comedy song "Gee, Officer Krupke," which occurs in the midst of the tragic events in Act Two of *West Side Story*. Arthur Laurents says he felt a need for comedic relief to lessen the tension and thereby increase the impact of the serious events to follow. Another example is "Master of the House" from Act One of *Les Misérables*; although the lyrics are not particularly witty in and of themselves, the song is funny because of its contrast to the grim tone of the rest of the show. In contrast, most of the score of *A Funny Thing Happened on the Way to the Forum* provides relief *from* the breakneck comedy of the scenes. Another example is the young soldier's ballad "Momma, Look Sharp" from *1776*, which is a total change of pace and viewpoint from the rest of the show. Some may find it superfluous, but for many it provides another perspective on the times, and a break from the central plot.

Songs may also be *diegetic*. This term comes from film, where diegetic or "source" music is what might be called "realistic" music—music that the characters in the film can hear, played either by musicians in the scene or from a recording. Diegetic film music stands in contrast to underscoring, which is implicitly accepted by the audience as a narrative, nonrealistic device to amplify its emotional response to the movie.

In modern musicals, most songs are considered to be nondiegetic, part of the narrative framework. When in *Oklahoma!* Curly sings "The Surrey with the Fringe on Top" to Laurey, we accept it as a nonrealistic but representational convention. Curly is trying to persuade Laurey to go on a date, but we don't really believe that in real life this uneducated cowboy could or would sing a song full of clever rhymes, catchy melody, and perfect eight-bar phrases, all accompanied by an invisible orchestra. But many musicals also contain presentational or diegetic songs. When characters are perceived by other characters (or by themselves) to be singing a song rather than talking, the song is diegetic. In *Kiss Me, Kate*, "We Open in Venice" is a diegetic song: The characters singing are professional performers, presenting a musical version of *The Taming of the Shrew*. Every show-within-a-show number is, by its very nature, diegetic.

The distinction between diegetic and nondiegetic songs is not black and white. Many theater songs are both diegetic *and* narrative in varying degrees, and musicals often take advantage of this overlapping of functions. For instance, at the beginning of Act Two of *Oklahoma!*, "The Farmer and the Cowman" is a song that Mr. Carnes sings during a social dance. But it is not strictly a "realistic," diegetic song, because Carnes gets interrupted by Will, Curly, and Ado Annie. We accept the convention that these three characters, none of them trained as poets, can improvise sung lines that rhyme and fit the phrases of the song perfectly. The line between diegetic and narrative song is blurred even further in the "Loveland" sequence near the end of *Follies*. We watch Ben, Buddy, Sally, Phyllis, and their younger selves performing numbers that are clearly rooted in the narrative; the numbers are not part of any "real" show, and as far as we know, neither Ben nor Buddy has ever performed before. However, the songs are performed as presentational, diegetic numbers, part of an imaginary "Follies" revue that seems to exist only in the characters' minds.

It is also important to note that the purely diegetic use of song in a musical is a very easy, very old device that can cheapen or ruin a show. Back in the 1920s, when the integration of songs into shows was much looser than it later became, a song might be introduced by one character saying to another, "Say, I really liked that song you sang yesterday—sing it for us!" Then the other character would step up and sing the song. But audiences today are usually too knowledgeable and sophisticated for that kind of shoehorning to work. A purely diegetic song truly stops the show, since it does nothing either to further the story or to reveal character, unless the way that the character performs the song reveals something about him. Except in revues, a song needs to be an integral part of a show, even if it is presented as a performance within the story.

So if a show-within-a-show song, or any diegetic or presentational number, is by definition not narrative in function, then how can it be integrated with the rest of a show? Probably the best solution has been to integrate it *thematically*, as in the concept musical. As mentioned in chapter 1, one feature of the concept musical is that the songs are usually not part of the narrative. Instead, these songs comment on the narrative, often from outside the scene. *Cabaret*, a classic concept musical, illustrates this technique. The numbers sung by the performers in the cabaret scenes may seem purely diegetic, but they are not simply a random group of songs. Each one comments on the scene that precedes or follows it. For instance, Scene Nine of Act One, in which Cliff agrees to do an unsavory errand for money, is followed by a cabaret scene in which the Emcee sings "The Money Song," about the country's desperate economic condition. Similarly, Scene Two of Act Two, in which we see the romance between Fräulein Schneider and Herr Schultz in jeopardy because he is Jewish, is followed by the cabaret song "If You Could See Her through My Eyes," in which the Emcee presents his Jewish sweetheart as a gorilla. In short, while the cabaret songs are diegetic, each is also integrated thematically with the rest of the show, and closely tied by subject matter to the scenes around it. If a show does call for a diegetic song, it works better when the song has strong connections with the rest of the show—even if they are subtle or subliminal. Such connections only increase the audience's enjoyment and enhance their experience of the show as a whole.

Finally, there are songs whose only purpose is to entertain. Many shows have one or more big production numbers that have little or nothing to do with the story, but exist to get the audience excited and applauding. Writers with aspirations to quality try to avoid these lowest-common-denominator songs, because they can indeed stop the show and tear the fabric of the story.

Certainly if a producer is paying for a large cast, that cast should be used and featured, and a big show should have some big numbers. Such numbers are problematic only when their raison d'être is to stun the audience with size or spectacle, rather than to advance the story in some way. Sometimes a show can get away with such divertissements, especially if the show is simply trying to please an undiscerning audience. The title song of *Hello, Dolly!*, for instance, is absurd and incongruous in its context. As William Goldman has pointed out, in a less superficial show this song would be a disaster, but in that show it is a highlight, especially with Gower Champion's staging. *Les Misérables* is more ambitious. But it ends with a full-throated reprise of an anthem of revolution by the entire cast, and though this reprise is rousing and moving to many people, it makes no sense and has no connection to the scene, which is about the protagonist Jean Valjean and the only two people that matter to him. This ending works only because of the size and aural impact of the forces involved, and because a big and solemn finale, even an absurd one, seems appropriate to such a big and solemn show.

SONG SPOTTING

Writers use the terms *song spotting* or *routining* to denote the process of choosing where to put the songs in a musical. This choice involves not only which places in the libretto should have music, but also what each number should be about, what its mood or style should be, which character(s) should sing it, and whether it should be a song, a dance, or both. Song spotting is an essential aspect of the libretto, and it is as necessary for a through-composed musical as for a traditional dialogue-and-song show. As with many other aspects of a show, song spotting choices made early often end up having to be changed: A song may be moved to a different spot, given to a different character, or replaced by another song, a dance, or nothing.

The first step in song spotting is for the writers to determine which places in the libretto should be musicalized. Most shows will have many different possibilities for songs. But good writers, consciously or not, follow the principle of musicalizing the emotional peaks of the story. If the story is at all suitable, there will be at least four or five such places. These are usually essential points of the central plot line; the songs in these places, if they are appropriate and well written, comprise the spine of the score, the essential songs. Among the most common locations for such high points are the opening of the show, the end of the first act, and the climax of the show in the second act, although the latter two locations often use reprises of earlier songs rather than new songs.

The climax of the show is sometimes the place for an "eleven o'clock number," so called because when evening performances of Broadway musicals used to start at 8:30 P.M., the climactic point would occur around 11:00. Even in this period eleven o'clock numbers were by no means universal, but they were often among the most memorable songs in a show. Classic eleven o'clock numbers include "Luck Be a Lady Tonight" from *Guys and Dolls*, "I've Grown Accustomed to Her Face" from *My Fair Lady*, "I'm Going Back" from *Bells Are Ringing*, "Rose's Turn" from *Gypsy*, and the title song from *Cabaret*.

Once the essential songs have been determined, the rest of the score can follow. These other numbers may be moving, funny, exciting, or full of surprising ideas that enhance the story, but they remain optional—no matter how good they may be, they can be dropped if necessary, and it is usually one of them that goes first if cuts or changes are made. For example, when *Fiddler on the Roof* was in rehearsal, there was a song in Act Two for Motel and Tzeitel called "Dear Sweet Sewing Machine." According to lyricist Sheldon Harnick, it was one of the most popular songs from the show in all the backer's auditions. When Jerome Robbins started rehearsing it, he felt there was something wrong, but could not put his finger on it—until the first out-of-town preview, when the song got no applause at all. No matter what the creative team did, it never got applause, so they eventually cut it. They later realized that by that point in the show, the audience was simply not interested in further exploring Motel and Tzeitel's relationship, which had been resolved in the first act.

Because few shows have fourteen or fifteen essential high points, there is no score without at least some optional numbers, and it can be instructive to analyze a show's score to distinguish the essential ones from the optional. For instance, in *The Music Man*, the category of essential numbers would include the song in which Harold Hill first mesmerizes the townspeople, "Ya Got Trouble." On the other hand, a number like "Shipoopi," in which Marcellus leads the children in a wacky dance, is clearly not essential; it is there as a production number, not a necessary point of the story. Another example of this distinction can be found in *A Little Night Music*. "Send in the Clowns" is an essential part of the story, while "The Miller's Son," though brilliant and effective, is not; it is sung by a minor character and has no important function in the main plot. All of this is not to say, of course, that such songs don't add a great deal to their shows—only that they're not essential. As Stephen Sondheim has said, "It's a question of how the dramatic arc of the show progresses and where the music is required."

Another question to be considered is how many numbers the show needs. While there are no rules about this, the vast majority of musicals have fourteen or fifteen songs in the score. To some extent this may depend on the nature of the show and the length of the songs. While *Pacific Overtures* has only eleven songs, for instance, most of them are unusually long.

Still another consideration is how much time there should be between songs. In a traditional book show, songs and scenes usually alternate. But many shows have two or even three numbers in a row, and many shows have a few scenes with no songs at all. In this area there are no immutable rules; the routining depends on the particular project. Still, two general principles apply to traditional shows. First, no matter where the songs are placed, the overall balance between music and dialogue is always about half-and-half. Second, the interval between musical numbers is never very long—perhaps ten minutes at most, or as a rough equivalent, one major scene plus one short one. While this may seem overly doctrinaire, it makes theatrical and psychological sense. If music takes less than half of the overall playing time, or if there are long intervals without music, the show will seem, if only subliminally, not "musical" enough to be a musical.

This problem does not arise in through-composed shows, of course. But even these shows are almost never routined as simply one song after another. Usually songs are separated by episodic material such as recitatives, interludes, short reprises, or even short spoken sections. Even when the music is virtually continuous, it seems that audiences need songs to be set off a little bit from each other. Otherwise, the show begins to seem like a concert or a revue—as some rock operas and poperettas do. Even *Cats* has episodic material scattered throughout—for example, the little spooky anticipations of "Macavity" at several points before the song itself is sung in Act Two, or the brief quotes of the "Jellicle Cats" theme throughout the show.

The last major consideration for the sequencing of the score is that of contrast or variety. A score needs not only variety overall, but contrast from one song to the next. This contrast can involve the number of people singing, the gender of the singer(s), the type of song (ballad, comedy song, and so on), the general emotional tone or mood of the song (happy, angry, sad, and so on), the musical style of the song, the meter, the tempo, the key, the harmonic mode (major or minor), and the amount of dance or choreographed movement. For instance, songwriters have traditionally tried to avoid having two or more ballads, duets, or big dance numbers in a row. Nowadays shows tend to be a little more flexible in this principle, but good writers are still mindful of the need for contrast. For example, if a good show has two ballads in a row, they are not both solos for male singers. This principle of contrast works a little differently for through-composed musicals, of course—but contrast is still necessary, or audiences may find the score monotonous.

We are not trying to give the impression that all musicals, or even all good ones, are created with the kind of logical forethought and planning that the preceding analysis implies. (We will discuss the actual steps of creating the libretto and score of a musical in part II.) But while many shows are not created with such conscious planning, the successful ones have addressed all of these issues.

Conventions of Musical Theater

Some of the conventions of musical theater, such as overtures, curtain call and exit music, dance music, and underscoring, will be covered in chapter 6. But there are two conventions that need to mentioned here. One is the *lead-in*. In a traditional show, this is the crucial moment immediately before a song starts, when a transition is made between spoken dialogue and song. (Some theater people call this moment a *gozinta*, because it "goes into" the song.) Lead-ins are among the most reworked parts of the libretto, because it is often difficult to find the right way to lead into a song. On the one hand, there must be some sort of transition, or the onset of the music may seem jarring. On the other hand, there is always a danger of saying too much, of "giving away" the essence of the song, which can make the song itself seem unnecessary or redundant. A lead-in is a place where songwriters can make contributions to the dialogue, because they often have the clearest sense of what is needed to set up the song most effectively. A lead-in is often accompanied by soft underscoring, which imperceptibly turns into the accompaniment of the verse or chorus of the song. Through-composed shows often have no lead-ins per se, but they still often need transitions from speech or recitative to a song. Even concept shows, which eschew smooth transitions, need some convention or technique to move between scenes and songs.

The other important convention is the *reprise* (pronounced *re-PREEZ*), the reiteration of any song first heard earlier in the show. A song can be reprised either in whole or in part; since the first appearance of a song often involves verses and interludes, partial reprises are far more common. The reprise can also be either literal, with the same lyrics as in its first appearance, or altered, with new lyrics for a different situation. The reprise used to be very common for one reason: It gave the songwriters additional chances to present the tunes they hoped would become popular. Nowadays, since musical theater songs rarely become popular hits, this is not so important. In addition, the greater sophistication of many contemporary musicals makes it inappropriate to reprise a song unless it serves a dramatic function. But there are many reprises that do make an emotional impact by recalling an earlier moment or emotion in another context. As an example, in the next-to-last scene of *Brigadoon* Tommy sits while his fiancée, with whom he is no longer in love, chatters away. Her words keep triggering memories of Brigadoon, each of which is represented by a partial reprise of an earlier number: "Come to Me, Bend to Me," "The Heather on the Hill," "I'll Go Home with Bonnie Jean," "From This Day On," and "Down on MacConnachy Square." Although this scene is somewhat excessive, it unquestionably makes effective use of reprises, and they serve an essential plot point: They lead Tommy to decide that he must return to Brigadoon, thus resolving the essential conflict of the show.

Stephen Sondheim once said he doubted that literal reprises could ever be justified; between an early scene and a later scene a character changes, and so the lyrics should change as well. He mentioned that while working on *Do I Hear a Waltz?* Richard Rodgers wanted a ballad to be reprised. When Sondheim asked why, Rodgers said he wanted the audience to hear the tune again, which Sondheim found an insufficient reason.

If a reprise uses altered lyrics, it can often be dramatically effective to change some aspect of the music as well—the person singing, the musical style, the tempo, the rhythm, or the instrumentation. In *A Funny Thing Happened on the Way to the Forum*, Sondheim creates comedy with a reprise by changing the gender of a singer. "Lovely" is originally sung by a boy and a girl, but its reprise is sung by two men, one dressed up as a girl. In its first appearance Hero and Philia are falling in love, while in its reprise Pseudolus is trying to persuade Hysterium that in women's attire Hysterium looks lovely enough to make people believe he is a girl. It is even possible to change harmonies in a reprise—for instance, an originally diatonic progression could be reprised with more chromatic harmonies to give it a poignant, bittersweet quality.

Many through-composed shows, in addition to ordinary reprises, make extensive use of another kind. Melodies in shows by Andrew Lloyd Webber or Boublil and Schönberg are frequently reprised, in whole or in part, throughout the show. Some writers consider this to be a variant of the *leitmotif* technique developed by Wagner for his later operas. A leitmotif is a musical motif or theme that is associated with a character, an object, an emotion, or a situation; it is repeated, transformed, and developed to reflect the journey of the character or object throughout the course of the drama. The technique used by Lloyd Webber and his imitators, however, bears little resemblance to the sophistication and power of Wagner's technique. First, these shows do not vary and develop short motifs or mutable themes, but reprise songs or large sections of them again and again with virtually no change—in *Phantom of the Opera*, for instance, the "Angel of Music" tune or the theme of "Masquerade." Second, the reprises are used indiscriminately and without dramatic logic or consistency. In *Phantom*, there are dozens of instances of a song originally sung by one character being reprised in a totally different context by another. For example, the "Masquerade" theme is the tune played by the Phantom's music box at the beginning and throughout the show, but it is also used as an underscore during the love scene between Christine and Raoul at the end of Act One; and it is sung by the opera company at a party in the first scene of Act Two. Neither of the latter two situations are directly connected with the Phantom. Similarly, the final cadence of "The Music of the Night," the Phantom's big ballad, is heard in the middle of the introduction to "All I Ask of You," which is Raoul and Christine's love duet and likewise has nothing to do with the Phantom. Indeed, the last fifteen minutes or so of almost every Lloyd Webber show consist largely of reprises strung awkwardly together. These indiscriminate reprises make no dramatic sense, and seem to be there only to drum the tunes into the audience's ears.

Establishing Time and Place

In a good show, the dialogue and lyrics seem appropriate for the time and place of the show, while remaining comprehensible to a contemporary audience. This is not always easy to do, especially when a show is set in a distant time and place. Past a certain point, however, this task becomes easier rather than harder. For instance, we have little idea how people in ancient Mongolia talked, so to put their dialogue into modern English might work, as long as clearly anachronistic words and slang were avoided. We have a much better idea, however, of how people in late-sixteenth-century England talked: They spoke an earlier version of English. So should one write dialogue for a show set in sixteenth-century England using their language? In that case, it would be hard for modern audiences to understand. (So is much of Shakespeare, of course—but audiences don't expect to hear Elizabethan English in a modern musical.) Or should one use contemporary English? In that case, it wouldn't have much flavor of sixteenth-century England. Striking the right balance can be difficult.

Questions of foreign or archaic language and usage demand knowledge and taste from the lyricist and librettist, but some general principles apply. The first and most obvious is that a commercial musical must be written in the language of the audience. Shows written for production in America or England are written in English, regardless of their settings. They may be set in Japan, France, Austria, or a Jewish *shtetl* in Russia, but that doesn't mean that their dialogue should be in Japanese, French, German, or Yiddish. *Pacific Overtures, La Cage aux Folles, The Sound of Music*, and *Fiddler on the Roof* were all written in English.

Since the presumption in English-language plays, musicals, movies, and books set in foreign countries is that the English dialogue and lyrics are "translations" of the language the characters would have actually spoken, it makes no sense for certain words or phrases to be left in the original language. If the English spoken by Parisian characters in a show is supposed to be French, then it would be absurd for one of them to break out with "Sacre bleu!" In 1953 Cole Porter got away with having Parisians sing *"C'est magnifique"* in *Can-Can*, but it would be much harder to get away with such a solecism today. The book and lyrics of *Nine* are often jarring because they are peppered with words and phrases in Italian, even though the characters are supposed to be speaking Italian throughout.

Fiddler on the Roof provides a better model. This show, set in a Jewish village in czarist Russia, is entirely in English, except for a few words of Russian (sung by a Russian during the song "To Life") and three Hebrew phrases (*l'chaim, mazel tov,* and *amen*). None of these violates the rule, because the everyday language of these people is Yiddish. The convention of *Fiddler* is that Yiddish has been made equivalent to, and translated into, English. It follows that Hebrew and Russian, being "foreign" languages, have not. The "foreign" terms are used sparingly, and in contexts that make their meanings clear. For instance, the first appearance of the Hebrew *"l'chaim"* is introduced by its English translation, "To Life."

The question remains, though, how writers may give the flavor of another time, place, and language when writing in English for a contemporary audience. Sometimes they ignore the issue entirely. As we indicated above, this can work for shows set sufficiently far in the past, or in mythological or legendary times, such as *Camelot* or the first act of *The Apple Tree*. On the other hand, if the original language is English and the time is not too far in the past, actual usage can be employed without the language becoming incomprehensible. Oscar Hammerstein II was a master of this, with expressions like "Foot!" in *Oklahoma!* or " 'D ruther not say" in *Carousel*. Sometimes, when English is standing in for another language, no regionalisms are needed, as in the turn-of-the-century "Swedish" of *A Little Night Music*.

A few well-placed exotic phrases, real or invented, often do wonders in suggesting a different setting when added to standard English usage. Frank Loesser sets the Broadway tinhorn-gambler milieu at the beginning of *Guys and Dolls* in "Fugue for Tinhorns" with phrases like "bum steer," "the morning line" and "feed box noise." In *West Side Story*, the atmosphere of 1950s youth gangs and slang is established in Arthur Laurents's book with such invented phrases as "Cut the frabbajabba," "rev us off," and "cracko, jacko," and Stephen Sondheim follows suit with phrases like "top cat in town" and "the swingin'est thing."

Theater is an illusion of life, not a reproduction. Writers who want exotic flavor generally realize that, as with any strong spice, a little goes a long way.

SCRIPT CONVENTIONS

At the end of this chapter are two sample pages that show the typical format for the libretto of a new show. This could be called the basic "playscript" format, because it is universally employed for the scripts of both new unproduced shows and shows in rehearsal, either before or after the show has opened.

Sample page A shows many of the conventions used in a traditional playscript, using a scene from an imaginary musical. These conventions have been around since the days when scripts were printed on manual typewriters. The headings denoting the act and scene, and the names of speaking or singing characters, are entirely capitalized, and tabbed so that they appear to be centered. Stage directions, which are also tabbed but not as much, are enclosed in parentheses. Ordinary speech is aligned to the left margin, with no first-line indentation. Lyrics are tabbed slightly and entirely capitalized, and usually prefaced with the stage direction "sings" or "singing." The capitalization and stage direction help to distinguish sung lyrics from spoken poetry or rhyming dialogue. Emphasized words or phrases are underlined.

Sample page B shows the same scene in a more modern version of script format, using the improvements that have been made possible by word processors and computers. On this page, the act and scene headings and the names of characters speaking or singing are truly centered, as are the stage directions. In the dialogue and lyrics, emphasized words or phrases are italicized rather than underlined. The stage directions are in parentheses and italicized, to distinguish them clearly from either parenthetical or italicized dialogue. (In some scripts, stage directions of paragraph length are not centered but tabbed, as in the older format.) Finally, the use of a proportional font rather than a monospaced font takes fewer pages and is more pleasant to read. With computers widely available, it would be foolish not to make use of the clarity and visual variety that word processors can give to a playscript.

Although the new script format approaches the look of a published book, both old and new versions differ in several ways from the format of a published libretto. Besides the paper size, the most noticeable difference is that lyrics in published librettos have traditionally been capitalized only at the beginning of each line and each sentence, like traditional verse. In the last few decades, though, some published librettos have begun to imitate the all-capitals format, which is also common for the lyrics of popular songs.

As with other technical aspects of the show such as lighting and props, most librettos specify little in the way of stage direction. Many directors and actors hate to see detailed directions in a script, feeling that it constrains them. Apparent exceptions to this can be found in the "acting editions" of plays and musical librettos from specialty publishers such as Samuel French and Dramatists Play Service. Acting editions are often full of cryptic, acronymic stage directions—like "He crosses S.R.C., picks up gun U.L., crosses D.C."—and many have set diagrams, property plots, and costume plots in the back. But these editions are published after the plays have appeared in commercial productions, and their stage directions and diagrams are from the original production, as an aid for productions by stock and amateur groups. They are in no way typical of ordinary script formats.

We would like to add a few observations about the length of scripts and the length of shows. Modern musicals have an average playing time of around two and a half hours, including a fifteen- to twenty-minute intermission. Writers can get a sense of the playing time of their show from the length of the script. With a 12-point monospaced font (in which all letters have the same width, as on a typewriter) and margins of one inch, the complete script of a libretto usually takes about 90 pages of letter-size ($8\frac{1}{2}''\times 11''$) paper. (Proportional fonts, like the one used in Sample page B, usually take fewer pages.) Of course this can vary to some extent; a fast-paced show, or one whose script has detailed notes on set or production elements, could take 120 pages for a two-hour playing time. But a 70-page script will probably run rather short, and a 150-page script rather long.

Shows can run shorter or longer than the average, of course. But if a Broadway show runs much less than an hour and a half, audience members may feel cheated if they paid full price; and if it runs longer than three hours, the producer must pay overtime to the musicians and stagehands, and most producers will not consider a show that entails this added expense. On the other hand, *Joseph and the Amazing Technicolor Dreamcoat* runs less than an hour and a half, yet it has had two successful Broadway productions; and *Les Misérables* ran almost three and a half hours when it opened, yet it was one of the longest-running shows in Broadway history. It should be added, however, that on Broadway *Joseph* added an elaborate, sung curtain call that was almost fifteen minutes long, and that *Les Misérables* had the insurance of having been a hit in Paris and London before it came to Broadway. In addition, many people found it unendurably long, and during the course of its Broadway run it was shortened by almost half an hour. Writers generally try to stay within these extremes.

Sample Script Page A (traditional format)

SCENE FIVE

(The park, evening. Jim and Janie enter.)

                        JIM

But Janie, I'm confused. I was trying to save the bandstand for the town. I didn't think you cared about the
money. If that's the way you feel, then why didn't
you just—

                        JANIE
          (leading him toward the bandstand)
Oh Jim, don't you understand? It's not just the money...
          (sings)

  THERE'S A THRILL YOU GET
  WHEN THE GRASS IS WET,
  AND THE GRASSHOPPERS JUMP AND SING.
  WITH THE BREEZE IN YOUR HAIR
  YOU CAN FEEL IN THE AIR
  THAT IT'S SPRING, SPRING, SPRING!

                        JIM
          (excitedly picking up his sousaphone)
You know, Janie, you're absolutely right! I can feel
it now myself!
                    (He takes her hand and they run up on the
                    bandstand, Jim dragging the sousaphone.
                    They ring the bells and bang the drums.
                    The other townspeople run in.)

                    TOWNSPEOPLE
          (variously, ad lib)
What is it, Jim? What's goin' on? Hey, what's the
ruckus fer?

                        JIM
Folks, I have an important announcement. The train <u>is</u>
going to come through town next month! And <u>I'm</u> <u>gonna</u>
<u>be</u> <u>on</u> <u>it!</u>
          (The crowd cheers.)

                        ALL
  WITH THE BREEZE IN YOUR HAIR
  YOU CAN FEEL IN THE AIR
  THAT IT'S SPRING, SPRING, SPRING!

          (The curtain falls quickly.)
                  END ACT ONE

Sample Script Page B (contemporary format)

SCENE FIVE
(The park, evening. Jim and Janie enter.)

JIM

But Janie, I'm confused. I was trying to save the bandstand for the town. I didn't think you cared about the money. If that's the way you feel, then why didn't you just—

JANIE
(leading him toward the bandstand)
Oh Jim, don't you understand? It's not just the money...
(sings)

THERE'S A THRILL YOU GET
WHEN THE GRASS IS WET,
AND THE GRASSHOPPERS JUMP AND SING.
WITH THE BREEZE IN YOUR HAIR
YOU CAN FEEL IN THE AIR
THAT IT'S SPRING, SPRING, SPRING!

JIM
(excitedly picking up his sousaphone)
You know, Janie, you're absolutely right! I can feel it now myself!
*(He takes her hand and they run up on the bandstand, Jim dragging
the sousaphone. They ring the bells and bang the drums.
The other townspeople run in.)*

TOWNSPEOPLE
(variously, ad lib)
What is it, Jim? What's goin' on? Hey, what's the ruckus fer?

JIM

Folks, I have an important announcement. The train *is* going to come through town next month! And *I'm gonna be on it!*
(The crowd cheers.)

ALL

WITH THE BREEZE IN YOUR HAIR
YOU CAN FEEL IN THE AIR
THAT IT'S SPRING, SPRING, SPRING!

(The curtain falls quickly.)

END ACT ONE

CHAPTER 4

THE CHARACTERS

Establishing Characters

CAST SIZE

The number of people in the cast of a musical can be as small as one or as large as one hundred, if the budget allows. There have been one-person shows, and the cast of *I Do! I Do!* consists of one man and one woman.

When writing for commercial theater, writers need to keep commercial realities in mind. To a producer, each role played by a separate actor means another salary to pay. Producers and writers must balance the logistical aspects of cast size—the size of the budget and of the intended theater—with the needs of the story.

In some instances the creative team decides that an actor will play more than one role. *Little Me* was written to showcase comedian Sid Caesar's talents, and he played the seven different men in the leading lady's life. In *Sunday in the Park with George* one actor plays both the impressionist artist George and his great-grandson, a multimedia artist also named George. The dramatic advantages in this case are clear, because we see the parallels between one George's artistic dilemma and the other's.

Sometimes actors double up on roles because there is not enough stage space to hold all of the performers that would otherwise be called for, or because the budget doesn't allow for additional actors. Small stock or touring productions tend to require that actors play multiple roles.

MAIN CHARACTERS

In theater, the main characters are known as the *protagonists* or *leads*. These are the characters around whom all of the onstage activity revolves. We learn more about the protagonists than we do about any other characters—they are presented whole from the beginning—and if the show works, we learn to root for them. There are invariably few lead roles in a musical. While there are practical reasons for this, it also allows the audience to maintain its focus on the main story with a minimum of distraction and confusion. Even ensemble shows like *A Chorus Line* tend to concentrate on a few characters.

It is a general rule that the protagonists of a musical must sing, or at least speak-sing, as Rex Harrison did playing Henry Higgins in *My Fair Lady*. Songs are the primary outlet for people's emotions in a musical; a main character who does not sing at all would not connect emotionally with an audience, which would negate the whole point of writing the show. Apparent exceptions to this, such as Susan the Silent in *Finian's Rainbow*, are secondary rather than main characters.

SECONDARY CHARACTERS

Secondary characters, while they are not as fleshed out as main characters, serve a variety of functions: as foils to the main characters, as relief (comic or other-wise) for the main plot, or as commentators on the main action. Sometimes they help to intensify the motivation for a lead character's actions. In *My Fair Lady*, for example, Colonel Pickering acts as a sounding board for Henry Higgins. But by making a wager with Higgins about the future of Eliza the flower girl, Pickering also helps to provide the impetus for the main action of the show.

In chapter 3 we mentioned that secondary characters often enact a show's subplot, which has its own dramatic arc and resolution. Secondary characters may get their own songs to sing, like the stepsisters in *Cinderella* and Doolittle in *My Fair Lady*; or they may sing primarily in ensembles, like Colonel Pickering; or they may not sing at all.

Theoretically, any characters other than the protagonists are considered "secondary," but even here there is a distinction. A tertiary role—sometimes called a "walk-on" part—is usually played by a member of the ensemble, briefly stepping out from the anonymity of the group to take a functionary role such as a butler or a waitress. Often these members of the ensemble are understudies for the lead actors.

ENSEMBLES

The ensemble, or "chorus," is usually the largest group of onstage performers. With no specific roles other than occasional walk-ons, a show's ensemble takes on different identities according to the needs of the show. In *The Sound of Music*, for example, members of the ensemble play nuns, Captain von Trapp's neighbors, and contestants in the Festival Concert. *South Pacific* requires members of the ensemble to portray islanders, sailors, marines, and officers.

As the costs of producing a musical steadily rose during the course of the twentieth century, the size of ensembles slowly shrank. In the "golden age" of the 1940s and 1950s, it was not uncommon for there to be both a singing chorus and a dancing ensemble with eight to twelve members each, usually divided evenly between men and women. By the 1950s even this was too large an ensemble for many shows. Just as leading performers were expected to be equally adept at lyricism and comedy, ensemble members found themselves expected to be "double threats," skilled as both singers and dancers—and often as actors as well, or "triple threats." In addition, with the disappearance of the obligatory opening chorus of "merry villagers" after *Oklahoma!*, writers no longer considered it a requirement that every show have a balanced ensemble of men and women. For instance, *A Funny Thing Happened on the Way to the Forum* has an ensemble of three men (the Proteans) and six women (the courtesans), while *Man of La Mancha* has thirteen men (the muleteers, Inquisition guards, and so on) and only two women, both of whom play small roles. Some shows require no ensemble at all. Writers often do not indicate the size of the ensemble, leaving it up to the director and producers.

Bringing Characters to Life

When actors prepare for roles, they usually try to learn as much as they can about the characters they are to play: not only basic physical aspects of the character like gender and age, but also the character's personal history. This sort of preparation gives actors insight into how their characters think and behave, and helps them to inhabit their roles with believability. Actors can find, or infer, much of what they need from the libretto and lyrics of the show. If that is not enough, they can get more from the source material, from research, or from their own imaginations. Thus they can reconstruct how their characters have lived up to the moment they appear onstage, and how they might behave as the play unfolds.

Often a show's creators will write character biographies or profiles themselves. The information can come from the original source, or the creators can use their imaginations to make up their characters' "backstories," as actors do. While these profiles usually do not appear in a libretto, they can inform a character's intentions, actions, reactions, and speech. A writer's profiles often include information about the characters' life histories, their physical types and vocal ranges, and their psychological and emotional makeup.

PHYSICAL TYPES AND VOCAL RANGES

Scripts usually contain little or no physical character description unless it is specifically needed for the story. The most basic physical aspects are gender, approximate age, and vocal range. The role may be male or female—but could it be either? Is the character young, old, teenage, or of indeterminate age, or does it matter? *A Funny Thing Happened on the Way to the Forum* originally called for Pseudolus, the main character, to be played by a man, but Whoopi Goldberg also played the role.

If a character's age is given, it is usually described as a range of years (for example, "early 20s" or "60–75"). Occasionally a role will be age-specific. In *The Sound of Music*, for instance, Liesl sings of being "sixteen going on seventeen." There is, however, nothing to prevent someone older or even younger from playing the part if they can do so believably. Other physical factors, such as height, weight, or build, do not appear in the script unless they are relevant to the story.

After gender and age, the final important aspect of a character's physicality is vocal range. The male lead could be a tenor, a baritone, or a bass; the female lead could be a soprano, a mezzo-soprano, or an alto. For some roles the range may not matter, but sometimes the quality of the voice may be important to the role, or in relation to the other roles. In chapter 6 we will talk about basic vocal ranges, but here let's consider some of the singing styles found in musical theater.

Early musical theater grew in part out of opera and operetta and, as a result, it often required "legitimate" voices for lead roles. A "legit" singer has been classically trained to produce a full, free sound, with agility, power, and dynamic control throughout the vocal range. Most lead roles in shows from the early twentieth century require legit voices, which could be heard clearly throughout the theater.

In addition to the legit soprano, mezzo, and alto, there is another type of female voice often used in Broadway shows, called a *belt*. A belter doesn't require the vocal agility or free sound of a legit singer as much as she does sheer power. A healthy belt is one that uses a mix of full (or "chest") voice and "head" voice to produce a powerful, almost steely sound. (The terms in quotes refer to where the singer perceives the sound to originate.) A belt singer needs consistent strength throughout most of her range. In different eras Ethel Merman and Betty Buckley have each been considered the epitome of the Broadway belter.

Songwriters, like other theatergoers, have often been seduced by the power and intensity of the belt, but this can be risky. Truly great belt voices are rare; often they are limited in range, or ineffective at ballads and other soft songs. Women who belt without the proper technique are apt to experience poor vocal health as well, and they may not be able to sustain a heavy schedule of eight shows a week.

With the mid-1950s came the rise of rock 'n' roll. It was associated with amplified instruments like the electric guitar, and voices became dependent on amplification just to be heard over the din. Of course, singing into a microphone required far less power than legitimate singing did. Rock also allowed for singers who didn't have "pretty" voices. A performer no longer needed to be classically trained.

Many shows like *Aida*, *Chess*, *Hair*, and *Tommy* require singers who can sing rock, rhythm and blues, or other popular idioms. Rock singing can be considered in some ways the antithesis of legit singing: Many rock singers aim not for beauty of vocal sound but for raw, untrained, elemental power. Of course, they need to develop and maintain healthy voices as much as legit singers do, although many rock singers don't learn this until they have already damaged their voices. It should be noted that many rock singers have limited vocal ranges and flexibility, which composers often have to take into account.

Men also have the option of using *falsetto*—a technique of singing with the "head" voice to obtain notes above the normal range. Falsetto can carry connotations of doo-wop nostalgia, otherworldly strangeness, or effeminacy, but in certain contexts it can be extremely effective, as in Jean Valjean's prayer "Bring Him Home" from *Les Misérables*.

PSYCHOLOGICAL AND EMOTIONAL MAKEUP

For characters' actions, dialogue, and songs to make sense, the characters must have an internal logic. Every word or lyric put in a character's mouth, and every action she takes, must be true to her nature.

In real life, people sometimes do things that we would consider "out of character." In a show, however, out-of-character words or actions are disconcerting. We need to see a character behave consistently, even if a real person in the same situation might not. An audience wants to root for the main characters. If their actions are too predictable, the show becomes boring—but if they are too surprising, they can break our suspension of disbelief.

Consistency from character to character within a show is also important. If most of the characters are two-dimensional "cartoons," as in a farce, then a character with the emotional and psychological complexities of Hamlet will seem incongruous on the same stage. A character who behaves in a manner inconsistent with the rest of the show will seem to be in a different show. Unless this incongruity is deliberately used as a running gag, good shows avoid it.

Finally, the protagonists in a musical must be likeable and interesting. The audience must quickly come to care about them and care what happens to them. If not, then the audience may be bored during their songs, as any of us might be bored hearing at length about the feelings of people for whom we don't care. Without people to root for, the narrative momentum can too easily be broken for the audience by the interruptions of musical numbers, scene changes, and act breaks. If the audience doesn't like the main characters, it has no reason to stick with them for two and a half hours.

Another way to put it is that the main characters must have *charm*. They do not have to be thoroughly wonderful people, with winning smiles and captivating graces, but there must be something about them to which the audience responds. Even characters such as Henry Higgins of *My Fair Lady* and Momma Rose of *Gypsy* are charming in their own ways. All it takes is something appealing in the character, perhaps even despite him- or herself, or something with which the audience can identify. In Rose's case it is her fierce devotion to her children and to bettering their lot in life, and her unashamed frankness about it; in Higgins's case it is his paradoxical, childish naiveté and directness. All of the major characters in every successful musical are charming in some way—even Sweeney Todd.

As we discussed in relation to the "principle of opposition," characters who remain optimistic in the face of adversity—even if they are completely out of touch with reality, like Don Quixote—are often appealing and likeable for that reason. The same is true of characters who are direct and honest about who they are and what they want. Other examples of characters who win us over by their optimism, courage, or honesty include Porgy in *Porgy and Bess*, Mrs. Anna in *The King and I*, and of course Little Orphan Annie in *Annie*.

Speech Patterns, Grammar, Vocabulary, and Dialect

We are each the product not only of our genetic makeup but of our education and of every event we have ever experienced. We use words we have learned and expressions with which we are comfortable; we say them using the speech patterns we've developed over time. We speak with accents or dialects. The way we speak is highly individual, even idiosyncratic.

In *My Fair Lady*, Henry Higgins talks of being able to pinpoint a man's background just by listening to him speak a few words, but we all learn a great deal about characters from the way they speak. Higgins himself has an upper-class British accent, and speaks with precise diction, correct grammar, and an extensive vocabulary. Everything he says, and the way he says it, indicates a high level of education and, by implication, a life of privilege and comfort. He speaks quickly and brusquely. His manner of speech shows an extraordinary amount of self-confidence, a lack of patience with the niceties of social interaction, and a fondness for the sound of his own voice.

Eliza Doolittle's speech, on the other hand, is often ungrammatical, with a smaller vocabulary, and with Cockney elisions mangling the words she does know. When upset she yells or makes ugly noises. Her speech shows a lack of education and implies a background of poverty.

Other patterns of speech can be influenced by psychological or physical problems such as a stutter, or Winthrop Paroo's lisp in *The Music Man*. Speech impediments are rare in musical theater because they can be a distraction from the rest of the show, or hinder the audience's ability to understand the words.

As an example of what dialogue can tell us about the characters speaking, let's consider an exchange from Act One, Scene Five, of *Camelot*, wherein King Arthur introduces Lancelot, a knight newly arrived from France, to his wife Guenevere.

ARTHUR

This is the Lancelot Merlyn spoke of. He's come all the way from France to become a Knight of the Round Table.

GUENEVERE

Welcome, Milord. I hope your journey was pleasant.

LANCELOT
(to Guenevere)

I am honored to be among you, Your Majesty. And allow me to pledge to Her Majesty my eternal dedication to this inspired cause.

GUENEVERE
(slightly startled)

Thank you, Milord.

(to Arthur)

How charming of you to join us, Arthur. This afternoon . . .

LANCELOT

This splendid dream *must* be made a universal reality!

GUENEVERE

Oh, absolutely. It really must. Can you stay for lunch, Arthur? We're planning . . .

LANCELOT

I have assured His Majesty that he may call upon me at any time to perform any deed, no matter the risk.

GUENEVERE

Thank you, Milord. That's most comforting. Arthur, we have . . .

LANCELOT

I am always on duty.

GUENEVERE

Yes, I can see that. Can you stay, Arthur?

ARTHUR

With pleasure, my love.
                    (He seats himself.)
I want you to hear the new plan we've been discussing. Explain it, Lancelot.

LANCELOT

To Her Majesty, Sire? Would Her Majesty not find the complicated affairs of chivalry rather tedious?

GUENEVERE
(frosting a bit)
Not at all, Milord. I have never found chivalry tedious . . . so far. May I remind you, Milord, that the Round Table happens to be my husband's idea.

This is not the Sir Lancelot of legends and old movies. This Lancelot is single-minded and perhaps a bit dense. He is educated and articulate, but he prattles, oblivious to Guenevere's signals that his audience has ended. We learn about Guenevere as well. She knows the rules of decorum and follows them as a queen must, even when presented with the exasperating blowhard that Lancelot appears to be. She is smart and witty, but she has a low tolerance for fools. King Arthur has only a few lines of dialogue, and as a result we don't get a lot of information about him. We can tell, however, that he is excited to have Lancelot join him in his cause, and we can guess that he knows to stay out of the way when Guenevere is irritated by the newcomer.

PRACTICAL CONSIDERATIONS

Performers have to rely on their voices constantly and consistently, and on Broadway they must do so eight times a week for several hours at a time. When a composer writes a song within a practical range, it puts a minimum of strain on the performers and enables them to pace their energy. A song written within a singer's normal range also affords her the opportunity to make the most out of her voice and put the song across more effectively. We will discuss vocal ranges in greater detail in chapter 6.

Sometimes a character requires a special skill, like riding a horse or bicycle, that can't be faked onstage. For instance, the title character in *The Will Rogers Follies* has to do tricks with a lasso. Good writers are aware of the practical difficulties in presenting such skills onstage. Singing while twirling a rope, for example, is not as difficult as singing while riding a horse, but singing while twirling a rope *and* riding a horse would be far more difficult. Writers of musical theater must always balance their desire to tell a believable and interesting story with the realities of theatrical presentation.

Chapter 5

The Lyrics

Before Writing

Musical theater is an art form that requires a thorough command of language in order to convey the thoughts and emotions of a variety of characters in a variety of situations. Before writing, lyricists must first know *how* to write. A lyricist doesn't have to be a great poet or playwright, or even have a great deal of training or education. But education and experience can only help. The better-read and educated lyricists are, the more varied and well-constructed their lyrics are likely to be. And the better their command of language, the greater the variety of people for whom they can write appropriate lyrics. A lyricist who stopped learning in high school might be able to write lyrics for Babe Ruth to sing, but not for Albert Einstein.

Terminology

Different people use different terms to describe the structure of a lyric. Here are some common terms and what they usually mean. A *line* is the equivalent of a complete or incomplete sentence. Regardless of its length, a line will consist of a pattern of stressed and unstressed syllables. These metric patterns are repeated in other lines. Two lines in immediate succession which share a metric pattern are called a *couplet*. Couplets usually rhyme (*a a*), though they don't have to (*a b*). A group of two or more lines in a pattern that will be repeated is a *stanza*.

(We follow the usual convention in which lowercase letters represent the metric pattern of, and the rhyme at the end of, particular lines; uppercase letters represent entire sections or stanzas.)

In various combinations, lines, couplets, and stanzas make up the sections of songs called verses, refrains, choruses, interludes, and codas or "tags." A song can be made up of any combination of these formal elements, which we will now define.

The term *verse* can denote several different things. Laypeople usually think of the verses of a song as different lyrics sung to the same music, as in "Old MacDonald Had a Farm," or "The Love of My Life" from *Brigadoon*. In these songs "first verse" means the lyrics that are sung the first time through the music, "second verse" means the lyrics sung the second time through, and so on. But for songwriters, and in this book, a verse is something different, namely the first or opening section of a lyric or song, which is usually subsidiary in some sense to the part of the song that follows.

The terms "refrain" and "chorus" have many different meanings to different people, but we will use them the way they are most commonly used by musical theater writers. A *refrain* is a recurring phrase, line, group of lines, or stanza. A *chorus*, on the other hand, is usually longer than a refrain, typically consisting of three or four stanzas or sections. It is the "song proper," the "body" of the song, as opposed to a verse or other introductory material. In many songs the chorus is repeated, usually with different lyrics—unlike refrains, which are usually repeated verbatim.

A refrain gives the essence of the song. It often contains the title; sometimes it is the title. In Lerner and Loewe's "Wouldn't It Be Loverly?" from *My Fair Lady*, each main stanza conjures up a picture of domestic bliss that is summed up by the refrain line at the end. On the other hand, in Adams and Strouse's nostalgic ballad "Once Upon a Time" from *All American*, each main stanza begins with the refrain line, which establishes that the scene of romance about to be described happened long ago.

Verses differ in length and structural significance. They are common in songs written before 1970, when rock music began to influence theater songs. When a singer performs the classic Gershwin song "The Man I Love" in its entirety, for instance, the first thing we hear is the verse, beginning with the words "When the mellow moon"; only after this introductory sixteen-bar section does the chorus begin with "Some day he'll come along." Similarly, the Rodgers and Hart song "My Funny Valentine" is introduced by the verse, which begins "Behold the way," before the chorus starts with the title line. In these old songs, the lyric and melody of the verse tend to be less interesting or memorable than those of the chorus, and to have no relation to the chorus. Not coincidentally, these verses are often omitted, especially in jazz performances.

As time went on, and Rodgers and Hammerstein ushered in the era of the integrated musical, the verse became more and more tightly woven into the fabric of the song—no longer poetically or musically negligible, and often repeated with different lyrics, like the chorus. In Act One of *Oklahoma!*, the first chorus of "People Will Say We're in Love," sung by Laurey, is introduced by a verse; when she finishes her chorus, Curly introduces his chorus with a verse of his own. (By the way, the melody of the verse, a downward scale—although unrelated to the melody of the chorus—is echoed in the orchestra after the first, second, and third lines of each "A" stanza of the chorus, which further helps to tie together the entire song.)

By the 1960s, songwriters had become even more creative with their verses. For instance, in *A Funny Thing Happened on the Way to the Forum* Hero's first song, "Love, I Hear," begins with a verse sung to the audience ("Now that we're alone") that establishes the naive, goofy, yet charming tone of the song and of his character. After the two choruses are over, there is a return to the beginning of the verse in the orchestra, followed by the last three lines of the verse sung with different lyrics ("Forgive me if I shout"), which give the song a feeling of ending as it began.

By the 1970s, verses as separate entities had largely disappeared from theater songs, except for intentionally old-fashioned ones like *Chicago*'s "When You're Good to Mama" or "A Little Bit of Good." If we look at the score to *Company*, for example, the only songs with sections that can be identified as verses are the title song and "Poor Baby," in both of which the verse is so tightly integrated that it cannot be removed without gutting the song.

A chorus is often sung several times in the course of the song. Take, for example, Ado Annie's song "I Cain't Say No" from *Oklahoma!* After an introductory verse, Annie sings a chorus (beginning with the refrain "I'm jist a girl who cain't say no"), then an interlude ("Whut you goin' to do"); then she sings a second chorus and a third, in each of which the refrain line is followed by new lyrics.

Types of Songs

There are many ways to classify songs. In this book we follow a classification commonly used by songwriters. It classifies songs by their *defining or essential element* as ballads, comedy songs, rhythm songs, charm songs, list songs, or musical scenes. Let's take a brief look at each type.

BALLADS

This category is probably the most familiar. Ballads are songs in moderate or slow tempo that convey a strong emotion through lyrical and expressive music. Melody and harmony are more important in ballads than in other types of songs, which is the main reason why composers like to write them. Another reason— that most hit songs from shows have been ballads—ceased to be true years ago.

Traditionally most ballads have been love songs, but after a century of musical theater it has become ever more difficult to find new words to express love. Coming out of the operetta tradition, Oscar Hammerstein II made a habit of putting twists on the traditional love lyric. When two young people meet in *Show Boat*, they don't say they've fallen in love, they say they could "Make Believe" they are in love. In *Oklahoma!* they joke that "People Will Say We're in Love," and in *Carousel* they imagine how it would be "If I Loved You." This tactic rapidly became overused, and lyricists found other ways to express love. In *Brigadoon* the course of romance between Tommy and Fiona can be charted by the titles of their ballads, from "The Heather on the Hill" (the beginning) through "Almost Like Being in Love" (not quite ready to admit it yet) through "There But for You Go I" (the impassioned avowal) to "From This Day On" (the farewell). By the 1950s the entire situation of the boy-girl romance was rather hackneyed, and love ballads had to become even more particularized and specific: "We Kiss in a Shadow," "I've Grown Accustomed to Her Face," "Small World."

Good musical theater always requires ballads, but good romantic love songs grow increasingly rare. In this respect, it is instructive to consider the ballads from *Fiddler on the Roof*. "Sabbath Prayer," "Sunrise, Sunset," and "Chavaleh" express the love of parents for their children; "Far from the Home I Love" is about the conflict between a girl's love for her family and her love for a man; and "Do You Love Me?" is a both a ballad and a comedy song, in which a couple acknowledges their love after twenty-five years of marriage. They are all about love, but none of them is a traditional boy-girl love song.

COMEDY SONGS

Comedy songs can be fast, slow, lyrical, or rhythmic; music is not the defining element. What is essential for a comedy song is that it be funny enough to make the audience laugh out loud. Experienced songwriters know that this is one of the hardest things to do well. Here are some famous comedy songs.

"Adelaide's Lament" *(Guys and Dolls)*
"Brush Up Your Shakespeare" *(Kiss Me, Kate)*

"Chrysanthemum Tea" *(Pacific Overtures)*
"The Company Way" *(How to Succeed in Business without Really Trying)*
"Dance: Ten, Looks: Three" *(A Chorus Line)*
"Gee, Officer Krupke" *(West Side Story)*
"A Hymn to Him" *(My Fair Lady)*
"I Cain't Say No" *(Oklahoma!)*
"When I'm Not Near the Girl I Love" *(Finian's Rainbow)*
"You Must Meet My Wife" *(A Little Night Music)*

Even this short list indicates the wide variety of ways in which a song can make people laugh. One conclusion that emerges from studying comedy songs is that good ones are based usually on character, occasionally on situation, but almost never on jokes. Another is how often the principle of opposition, discussed in chapter 3, works for comedy: In songs like "I Cain't Say No," "Adelaide's Lament," and "A Hymn to Him," the characters are frustrated or unhappy, and feeling sorry for themselves. Most comedy is based on people in unpleasant situations or sad moods.

RHYTHM SONGS

There are many songs in musicals in which the defining element is driving rhythm, especially big group numbers and dances. They are sometimes called "up-tempo" songs, although not every rhythm song is fast. These songs give musical theater much of its excitement and pizzazz. Here are some representative examples.

"Company" *(Company)*
"Get Me to the Church on Time" *(My Fair Lady)*
"The Jet Song" *(West Side Story)*
"June Is Bustin' Out All Over" *(Carousel)*
"Luck Be a Lady Tonight" *(Guys and Dolls)*
"Oklahoma" *(Oklahoma!)*
"Seventy-Six Trombones" *(The Music Man)*
"There Is Nothing Like a Dame" *(South Pacific)*
"They Call the Wind Maria" *(Paint Your Wagon)*
"To Life" *(Fiddler on the Roof)*
"You're a Good Man, Charlie Brown" *(You're a Good Man, Charlie Brown)*

CHARM SONGS

This category was named by Lehman Engel, and while not as well known as the others, it is an extremely valuable one. The defining element of a charm song is, of course, that it makes the character singing come across as charming. It is almost always in a moderate tempo, sung by a leading character. Through a good charm song, the audience comes to know and like the character. An archetypal charm song is Frank Loesser's "Once in Love with Amy" from *Where's Charley?*; when a performer truly has charm—like Ray Bolger, the original performer of this song—the audience will follow him anywhere and even sing along with him, as Bolger got audiences to do every night. Here are some more classic charm songs.

"Camelot" *(Camelot)*
"Getting to Know You" *(The King and I)*
"I Feel Pretty" *(West Side Story)*
"If I Only Had a Brain" *(The Wizard of Oz)*
"If I Were a Rich Man" *(Fiddler on the Roof)*
"I Got Plenty of Nuttin' " *(Porgy and Bess)*
"I Whistle a Happy Tune" *(The King and I)*
"She Loves Me" *(She Loves Me)*
"The Surrey with the Fringe on Top" *(Oklahoma!)*
"Wouldn't It Be Loverly?" *(My Fair Lady)*

LIST SONGS

List (or "laundry list") songs have no true progression or dramatic structure; they are lists of parallel things or ideas, which can usually be arranged in almost any order without hurting the song. A quintessential list song is Cole Porter's "You're the Top" from *Anything Goes*. No matter how you arrange the choruses of this song, or even some of the lines within stanzas, its message is unchanged. On the other hand, Porter's "It's Delovely" from *Red, Hot and Blue!*, which appears to be a list song, is not; it tells the story of a romance that leads to marriage and then children. Here are some other list songs.

"Ain't Got No" *(Hair)*
"It Ain't Necessarily So" *(Porgy and Bess)*
"My Favorite Things" *(The Sound of Music)*
"Tschaikowsky" (*Lady in the Dark*)

As successful as list songs were in the 1930s and 1940s, they have become rarer and rarer. Audiences have become more demanding, and they expect a theater song to do more than merely unreel in an amusing way.

Musical Scenes

This is a catch-all category that includes any musical number longer and more complex than a single song. Musical scenes usually incorporate all or parts of several different musical numbers. The classic example is "Soliloquy" from Rodgers and Hammerstein's *Carousel*. While it is considered a single number, it is actually a mosaic of different materials:

"I wonder what he'll think of me"	(incomplete song)
"My boy Bill"	(complete song)
"I don't give a damn what he does"	(interlude)
"My boy Bill"	(short reprise, with underscoring)
"And I'm damned if he'll marry"	(short interlude)
"I can see him"	(short interlude)
"I wonder what he'll think of me"	(short reprise, with underscoring)
"My little girl"	(complete song)
"I gotta get ready"	(final section or coda)

Other classic musical scenes include "I'm Going Back" from *Bells Are Ringing*; "I've Grown Accustomed to Her Face" from *My Fair Lady*; and "Rose's Turn" from *Gypsy*.

Song Structure

In terms of the arrangement of stanzas, a song's chorus can take various forms. The two most common are called ABAC and AABA; in these abbreviations, each capital letter stands for a stanza of lyrics and, usually, eight bars of music. The ABAC form is still occasionally used, although it is more typical of pre-golden age popular and theater songs. Here is a classic example, "A Pretty Girl Is Like a Melody" from Irving Berlin's score for the Ziegfeld *Follies of 1919*. Notice that the first and third stanzas (the A sections) are identical in metrical and rhyme schemes, while the second and fourth stanzas (B and C, respectively) are different from them and from each other.

A   A pretty girl is like a melody
    That haunts you night and day.

B   Just like the strain of a haunting refrain
    She'll start upon a marathon
    And run around your brain.

A   You can't escape, she's in your memory
     By morning, night, and noon.

C   She will leave you and then
     Come back again.
     A pretty girl is just like a pretty tune.

The AABA form, by far the most common form of the golden age, remains extremely common even today, not only in theater music but in popular ballads. Here is an example from *On Your Toes*, the first chorus of "There's a Small Hotel" by Lorenz Hart and Richard Rodgers. Note that the first, second, and last stanzas—the A sections—are identical in structure (except for an extra rhyme in the last stanza), while the third stanza, the B, is very different.

A   There's a small hotel
     With a wishing well.
     I wish that we were there
     Together.

A   There's a bridal suite,
     One room bright and neat,
     Complete for us to share
     Together.

B   Looking through the window you
     Can see a distant steeple.
     Not a sign of people.
     Who wants people?

A   When the steeple bell
     Says, "Good night, sleep well,"
     We'll thank the small hotel
     Together.

As important as AABA is, there are few lyrics in the pure form. Most of them are in a common variant usually called AABA', in which the final A or "A-prime" differs slightly from the first two. The difference may be a word or two, or a line, but sometimes an entire section is added at the end, called a *tag* or *coda*—the latter an Italian word meaning "tail." The final stanza of "There's a Small Hotel," which follows the chorus and an interlude, is an A':

A'   When the steeple bell
     Says, "Good night, sleep well,"
     We'll thank the small hotel,

We'll creep into our little shell,
And we will thank the small hotel,
Together.

A chorus almost always includes a contrasting *B section*. This section has two other common names: the *release*, because it serves as a release or relief from the A section; and the *bridge*, because it serves as a bridge between the second A section and the final one. It is not only useful but essential that the B section show a clear contrast in metrical and rhyming patterns (as well as in melody and harmony) to the A section. This allows a brief respite from the repetition of the A section, and instills a desire for its return, as the B section of "There's a Small Hotel" demonstrates.

A good release does more than change the metrical construction or rhyming pattern of a song; it also gives a shift in perspective, a new viewpoint, or a change of subject. In "There's a Small Hotel," the first two A sections narrow their focus from the hotel to the couple's room. The B section reverses this with a "wide-angle view" out the window, until the final A section brings the focus back to the couple.

It is also possible to write a type of AABA form in which each subsequent A, not just the last, differs slightly from the original A section. When this happens in good songs, the modification almost always appears in a single line toward the end of the stanza. It is almost never at the beginning of the stanza, nor in more than one place, because either arrangement might well make the stanza unrecognizable as a variant of the A section. Cole Porter was a master of this technique, of which "I Get a Kick Out of You" from *Anything Goes* is a perfect example.

A   I get no kick from champagne.
       Mere alcohol doesn't thrill me at all,
       So tell me why should it be true
       That I get a kick out of you.

A′   Some get a kick from cocaine.
       I'm sure that if I had even one sniff
       It would bore me terrific'ly too
       Yet I get a kick out of you.

B   I get a kick ev'ry time I see
       You're standing there before me.
       I get a kick tho' it's clear to me
       You obviously don't adore me.

A″   I get no kick in a plane.
       Flying too high with some guy in the sky

Is my idea of nothing to do,
Yet I get a kick out of you.

The first line of each A stanza is metrically identical and ends in an "ane" rhyme (*champagne/cocaine/plane*). The second line of each is metrically identical and contains an internal rhyme (*if/sniff, high/sky*, and so on). But in the second A section the third line has one more syllable than in the first A, and in addition there is now a third rhyme, *terrif-*(ic'lly), to go with *if* and *sniff* in the second line. While the last A has the same line lengths as the second, there is a new rhyme added to the middle of the second line (*guy*, rhyming with *high* and *sky*), while the fourth rhyme, *i-*(dea), is in a different place in the line from where *terrif-* was in the second stanza. (Porter also changes the melodic line.)

There are other song forms available, of course, that are unrelated to either ABAC or AABA structures. In *Finian's Rainbow*, for instance, "Look to the Rainbow" takes a verse-refrain or AB form. In this form, found in many folk songs, the verse is repeated with different lyrics and the refrain is repeated verbatim. The verse-refrain form is appropriate for "Look to the Rainbow," which has the flavor of an Irish folk song. On the other hand, "How Are Things in Glocca Morra?" has an AAB form (or in this case AA'B), sometimes called "bar form." Cole Porter also wrote songs in this form.

Theater songs use many other forms, but it would be difficult, and probably unworkable, to do without sections or stanzas altogether. The essence of every song form we've discussed so far is that some section—whether you call it A, B, or Z—is heard at least two or three times, so that the audience becomes familiar with its structure and melody; and it alternates with at least one section that is very different in lyric structure, point of view, melody, and often rhythm and harmony as well. As we will show in our next chapter, these aspects of good form are centuries old; they have lasted this long for reasons rooted in the way audiences hear and remember.

As always, there are exceptions. If any song is a classic theater ballad, Rodgers and Hammerstein's "You'll Never Walk Alone" from *Carousel* certainly qualifies. Yet the lyric of this song is in the form AABC, while the music takes the form ABCD! Still, an analysis will show that there are a number of ways, both lyrical and musical, in which the song is held together and logically organized. Also keep in mind that it was written by two of the greatest practitioners of the form, each of them with twenty years of experience and at the height of his powers.

A good piece of general advice for beginning lyricists would be not to start off by writing long or complicated forms. Mastery of any craft takes time, and the ability to handle complex forms and material takes experience. No one in the audience, except possibly other songwriters, gives any points for complexity. A good simple song is always better than a merely adequate complex one. In addition, the old ABAC, AABA, and other "32-bar" forms still have a vast amount of room for variety and novelty of thought and expression. At the turn of the twenty-first century, even such luminaries as Stephen Sondheim were still writing AABA songs.

We have not yet discussed the structure of verses, but usually they do have structure. They often use standard chorus structures in miniature. Here, for instance, is the verse of Irving Berlin's "A Pretty Girl Is Like a Melody." Silly as it now seems, it shows a perfect AABA form, except for one additional syllable in the second line of the first A.

A   I have an ear for music,
     And I have an eye for a maid.

A   I link a pretty girlie
     With each pretty tune that's played.

B   They go together,
     Like sunny weather
     Goes with the month of May.

A   I've studied girls and music,
     So I'm qualified to say . . .

Verses do not have to come at the beginning of songs; they can come between choruses as well, although then they are often called interludes. Kander and Ebb's "Cabaret" is an example.

Finally, let's consider a lyric not as a succession of formal structures, but as a *dramatic* structure. Returning to "I Get a Kick Out of You," we may note that as Porter's A section changes, it becomes not shorter or simpler, but longer and more complex. This intensification is part of the dramatic arc mentioned in chapter 3. Just as an arc of intensity rising to a climax, then falling in a denouement or coda, will be found in virtually every good book, play, or movie, so it can be found in most good songs as well.

There are, of course, good songs that do not have the dramatic arc. Many of them are list songs. "I Get a Kick Out of You" verges on being a list song, but as we have seen, Porter sets up an intensification that helps give the lyrics and music a "build." There are other intensifications as well. For instance, the chorus starts with a reference to champagne, which might be called a conventional thrill; proceeds to cocaine, a more exotic and illicit one, especially in 1934; and ends up high above the earth, both lyrically and musically.

Whether they tell a linear story or not, good lyrics always move in some logical way. Even if the audience doesn't notice, it still makes the song better; no one can see the girders in a skyscraper, but imagine how long it would stand without them. Even if a song is a simple ballad that barely travels from point A to point B, organization of some kind makes it that much tighter, that much stronger—and that much harder for potential producers to find fault with. When songwriters mix ideas or images of one kind with those of another, or put their ideas in a random order, the song can confuse the audience; or else it becomes a list song, which nowadays may bore the audience.

There are many ways to organize a lyric besides telling a story. Here are just a few examples. In Lerner and Loewe's "If Ever I Would Leave You" from *Camelot*, each stanza uses imagery from one of the seasons, in order: summer, autumn, winter, spring. In Stephen Sondheim's "Liaisons" from *A Little Night Music*, Madame Armfeldt recalls three past relationships: the first with a baron, the second with a duke, the third with a king. In Rodgers and Hammerstein's "Something Wonderful" from *The King and I*, the first stanza of the chorus deals with what the king says, the second with what he does, the bridge with his dreams, and the final stanza with how a woman who loves him is compelled to act. In Meredith Willson's "Till There Was You" from *The Music Man*, Marian mentions in each stanza something that she had never noticed until she met Harold Hill: In the first stanza it is bells ringing on the hill, in the second birds flying in the sky, in the bridge a number of different sensory images (music, roses, fragrant meadows), and in the last, love singing all around her.

Now let's take a look at the structural requirements of the different types of songs.

BALLAD STRUCTURE

The essential job of a ballad is to express emotion. As we have seen, the emotion does not have to be expressed immediately, directly, or even consciously. There are ballads, like "People Will Say We're in Love," where the emotion isn't explicit lyrically until the very end, though the music makes the emotion clear from the beginning. There are almost as many ways to structure ballad lyrics as there are ballads, but since they are about emotion, and since it is almost always more effective to build emotion over the course of the song, the dramatic arc is usually incorporated in some way. Let's take Sheldon Harnick and Jerry Bock's "Far from the Home I Love" from *Fiddler on the Roof* as an example.

This song occurs at one of the emotional crises of the character Hodel's life: She has been forced to choose between staying with her beloved parents or going into exile in Siberia to join the man she loves, and she has chosen the latter. Now she has to explain her choice to her father.

A   How can I hope to make you understand
     Why I do what I do,
     Why I must travel to a distant land,
     Far from the home I love?

A   Once I was happily content to be
     As I was, where I was,
     Close to the people who are close to me
     Here in the home I love.

B   Who could see that a man would come
     Who would change the shape of my dreams?
     Helpless now, I stand with him,
     Watching older dreams grow dim.

A   Oh, what a melancholy choice this is,
     Wanting home, wanting him,
     Closing my heart to every hope but his,
     Leaving the home I love.

A′   There where my heart has settled long ago
     I must go, I must go.
     Who could imagine I'd be wand'ring so
     Far from the home I love?
     Yet, there with my love, I'm home.

The first stanza expresses Hodel's problem. In the second stanza she begins trying to make her father understand her conflict; she assures him that when she was younger, she was happy at home. In the B section, though, things intensify; time has passed, she has grown older, and a man has entered her life with whom she has fallen in love. Now that she has reached the present, in the third A she states the conflict in its bare essentials ("Wanting home, wanting him"), and she makes clear that, having to make a choice, she has chosen to go with her lover. This is the dramatic climax of the song, but the story isn't over. Harnick adds one more A section—a final A′ in which Hodel reiterates that her mind is now made up and expresses wonder, if not regret, that she is leaving. Then there is a final twist: although she is leaving home, she will be home wherever her lover is. While a "twist ending" is in no way a necessity for a ballad, it does add a last poignant note, and many classic ballads have one. Incidentally, an AABAA' form might be expected to make a repetitious or awkward song, but Bock's musical setting avoids this, simply and brilliantly. We will discuss the musical structure of this song in the next chapter.

This short lyric expertly establishes a situation, builds to a climax, and leads to a moving conclusion. Almost every good ballad incorporates the dramatic arc in some simple, economical way.

COMEDY SONG STRUCTURE

With comedy songs the imperative is simpler, although often more difficult to execute: to make the audience laugh. All comedy songs can be reduced to one of two basic structures. Either the entire song is one long joke, like "Chrysanthemum Tea," or it is a series of jokes, like "Adelaide's Lament," "One Hundred Easy Ways to Lose a Man," or "Gee, Officer Krupke." (When song-writers use the term "joke," they don't mean the kind of joke with a punchline told by old-time stand-up comedians, but a line or stanza intended to make the audience laugh.) In either case, a similar structure is necessary. If it is a long joke, then the whole song must lead up to the payoff, which must occur either at or shortly before the end. If it is a series of jokes, then the biggest joke must come last, or the audience will be disappointed; the jokes must build as the song goes on, until the biggest joke occurs at the end. With either type, the writer must structure the entire song by working backwards from the end.

Comedy songs must also allow time for the audience to react. This sounds obvious, but it is harder in a song than in dialogue, where the actors can simply stop talking and wait until the laughter is over. Often the laugh lines occur before the end of the stanza, and the music can't just stop without destroying the fabric of the song; nor can the song proceed directly to the next laugh, or the audience will miss some of the setup and it will fall flat. Here is the first chorus of a comedy song that solves the problem brilliantly, "If Momma Was Married" by Stephen Sondheim and Jule Styne, from *Gypsy*.

A   If Momma was married we'd live in a house,
    As private as private can be:
    Just Momma, three ducks, five canaries, a mouse,
    Two monkeys, one father, six turtles and me . . .
    If Momma was married.

A   If Momma was married, I'd jump in the air
    And give all my toeshoes to you.
    I'd get all these hair ribbons out of my hair,

And once and for all, I'd get Momma out, too . . .
If Momma was married.

B   Momma, get out your white dress!
     You've done it before—
     Without much success—
     Momma, God speed and God bless,
     We're not keeping score—
     What's one more or less?

A′   Oh Momma, say yes
     And waltz down the aisle while you may.
     I'll gladly support you,
     I'll even escort you—
     And I'll gladly give you away!
     Oh Momma, get married today!

The lyric of this duet is structured so that the laughs, which gradually get bigger, are immediately followed by repetitions of the refrain, which the audience knows after hearing it once. Even the lines "Without much success" and "And I'll gladly give you away" have time to be laughed at, because each is followed by a long "Momma," which the audience doesn't need to hear again.

Songs with similar solutions include "Always True to You in My Fashion" from *Kiss Me, Kate*, and "The Little Things You Do Together" from *Company*. More than any other type of song, comedy songs are about the lyrics. The composer can write a good melody, but it is more important that the music stay out of the lyrics' way, letting the words register as clearly as possible to get their laughs.

RHYTHM SONG STRUCTURE

Even more than with ballads, there are so many ways to write rhythm songs that it is impossible to generalize about their shape. Still, the same basic rules apply; the more of a shape the song has, the more satisfying the audience will find it. For instance, the structure of *My Fair Lady*'s "Get Me to the Church on Time" is simple—a short verse and two AABA choruses, followed by a big dance; but even here there is a buildup of intensity, because Doolittle sings the first chorus by himself, and then the ensemble joins in the second chorus. After the dance, when the wild night is over and the sun is starting to rise, the song starts again, very slowly and quietly. Gradually it gets faster, and builds and builds, until at the very end it is even bigger and louder than before. The dance is the climax of the number, but this final chorus gives the entire number an exhilarating and memorable ending.

In contrast, consider "Luck Be a Lady Tonight" from *Guys and Dolls*. Here the structure is even simpler, just a verse and two AABA choruses. Again, though, a rise in intensity is built into the number. Sky Masterson sings the verse and the first chorus himself, then the ensemble joins him for the second chorus. Also, while most of the lyrics are the same in both choruses, the B sections are different. (The first begins "A lady doesn't leave," and the second "A lady wouldn't flirt.") There are three increases in intensity in the second B. First, an internal rhyme has been added to the third line (*make/snake*), giving it a little extra drive or push. Second, the ensemble starts to urge Sky to roll the dice during the same third line, adding to the intensity and, by implication, to the pressure Sky must be feeling. Finally, in the second B section Sky says he has bet his life on this roll—not only a sharp contrast to the cool, understated tone of the first B, but a frank confession. Musically, the entire second chorus is more intense than the first. Again, there is an increasing drive in the song that leads steadily to its climax, as Sky rolls the dice at the very end.

CHARM SONG STRUCTURE

There is no standard template for charm songs; they can be in any of the forms we have already discussed. The principle of making each song a journey is not as essential for a charm song as for a ballad or even a comedy song. Most charm songs do include a journey, but it is usually a short and gentle one. In today's ungentle age, though, it is increasingly difficult to find a charm song of any kind, or even to find characters for whom charm songs can be written. This may be one of the reasons why so few new shows have been successful in the last several decades, why so few of the successful ones have been good, and why so many recent Broadway productions have been revivals of classic shows.

As Momma Rose and Henry Higgins demonstrate, a character doesn't have to be cute and loveable in order to charm the audience. Neither *Gypsy*'s "You'll Never Get Away from Me" nor *My Fair Lady*'s "I'm an Ordinary Man" is a charm song in the "Once in Love with Amy" manner, and yet both accomplish the goal of making the characters that sing them—who are obnoxious and even monstrous—charming in some way. To write a charm song for a character, songwriters must find something in the character that is likeable—even if it is simple frankness or optimism.

A NOTE ON MUSICAL SCENES

Musical scenes must, by definition, have not only a clear dramatic arc but also one or more significant actions or moments (what actors call *beats*). The actual forms of the component parts can be almost anything, as our look at "Soliloquy" has shown—songs, pieces of songs, underscoring, reprises or partial reprises, or recitative. It takes time, and often more than one try, for writers to find the most effective way to set the scene to music. The actions to be set to music, however, must always be clear.

One final point needs to be made about theater lyrics. As songwriters write a song, they are the dramatists of that three-minute period of time, and they must conceive every song dramatically and theatrically. They are writing something to be acted, which means that some kind of action is taking place. Therefore, as the dramatists, it is their responsibility to have in mind what will actually be happening onstage while the song is being sung.

Once in a great while it is enough for a character to simply stand there and sing a song, but most of the time it will not hold the audience. If the songwriters have not thought out the song's staging, and the director cannot come up with a brilliant idea on how to stage it, there will be a big problem. This is not necessarily a reflection on the director's ability: Bock and Harnick wrote a song for *Fiorello!* that George Abbott, the most experienced and successful director in the business, couldn't find a way to stage. It was dropped from the show. Similarly, Jerome Robbins was staging *West Side Story* when he came to a pause in the middle of Tony's ballad "Maria." "What am I supposed to do there?" he asked Stephen Sondheim, the lyricist. "You've got to give me something for him to do." From this Sondheim learned that writers need to stage every song in their heads. Directors may then change anything or everything, but at least they have something from which to build.

Titles

Who cares about song titles when you're in the theater and the lights are too low to read the program? No one announces the title of a song during a musical. Still, song titles are of great importance to the lyricist and composer. The title of a song is usually fundamental; it embodies the central concept of the song. More often than not the title is the refrain, the phrase that keeps coming back and that the listener both expects and wants to hear. A glance at the song titles below will illustrate their importance within their respective songs.

"Being Alive" *(Company)*
"How Are Things in Glocca Morra?" *(Finian's Rainbow)*
"Put On a Happy Face" *(Bye Bye Birdie)*
"Tomorrow" *(Annie)*
"Tradition" *(Fiddler on the Roof)*
"With a Little Bit of Luck" *(My Fair Lady)*

Finding the right title for a song is one of the most crucial choices a lyricist makes, because the title sums up the entire song, and because as a refrain it often dictates the song's structure to a large extent. The title can come from the situation, the character, or the central theme or concept of the show. Here are three examples. Toward the end of *Bells Are Ringing*, a modern version of the Cinderella story, Ella leaves her Prince Charming's party, feeling that she cannot keep up the pretense of her Melisande identity. She sadly decides that she must give him up, and sings a ballad entitled "The Party's Over." In the first act of *The Apple Tree*, mother Eve has just had a baby, the first human baby in the world. She sings it to sleep with a lullaby—but because she doesn't know what a baby is, the lullaby is called "Go to Sleep, Whatever You Are." And in the comic *Little Shop of Horrors*, the nebbishy hero Seymour and his lady love Audrey sing a duet that manages to be a big love ballad while simultaneously kidding itself, with the title "Suddenly Seymour."

Song titles can appear anywhere in a song, but most often they are found in the first or last line of an A section of the chorus. Since the B section takes a different tack from the A sections, titles almost never appear there.

Some song titles do not appear *anywhere* in the song. This is especially true of comedy songs in which using the refrain as a title would give away the song's punch line. An example is the song sung by Val in *A Chorus Line* about her cosmetic surgery. The title, "Dance: Ten, Looks: Three," is taken from the first line of the opening verse, and it helps to set up the listener for what is to follow. But more to the point, this title saves the actual refrain ("Tits and ass") for a surprise.

Rhyme

One of the most important technical aspects of theater lyrics is rhyme. But before we get to the *use* of rhyme, let's first discuss the *purpose* of rhyme.

In most popular songs, drums and other percussion instruments are used to accent certain notes or beats in the music, as well as to help keep the beat. These accents point up certain words or phrases, or add syncopation and excitement to a phrase. They are not random; if they were, the result would be not exciting but chaotic. In a lyric, rhyme performs a similar function. In a good lyric, rhymes are not simply decorations applied to certain lines like racing stripes. They are used to point up important words or ideas, to clarify parallels or antitheses, or to make a line flow, soar, or even jump out at the listener. As a corollary, putting rhyme where it doesn't belong pulls focus away from the important things toward the unimportant, such as the lyricist's cleverness rather than the character and the story. As Stephen Sondheim has said, since a rhyme points up the word that rhymes, "if you don't want that word to be the most important in the line, don't rhyme it."

Good lyricists try not to rhyme except where there is a reason to rhyme. What's important is the thought or emotion the character is expressing. No matter how clever a rhyme may be, it is not an adequate substitute for a thought or emotion. There is nothing wrong with jotting down in advance rhymes that might be useful in a song, as long as these rhymes don't substitute for having something to say. Rather than trying to force a particular rhyme into a song, good songwriters first decide what they want to say, and *then* find a rhyme that helps make the point. The intelligent use of rhyme is considered a sign of good craftsmanship.

Sondheim has also observed that the more rhymes a song has, and the more elaborate and dense the rhymes, the more educated the character singing seems to be. Of course, this is not appropriate for every character, as Sondheim him-self came to feel about his lyric for "I Feel Pretty" in *West Side Story*; he has said that during this song Maria, an uneducated immigrant, sounds as though she would fit in perfectly at one of Noel Coward's parties.

In musical theater there is only one type of rhyme. A rhyme, or "perfect" rhyme, exists between two words or phrases when the initial consonants of the last accented syllables of both words or phrases are different, and the sounds that follow those consonants are identical. For example, "moon" rhymes with "soon," "sailor" rhymes with "tailor," and "bicycle" rhymes with "tricycle." Rhymes can be of any length, but most are one, two, or three syllables long.

An *identity* is similar to a rhyme, but in an identity the initial consonants of the last accented syllables are the same. (It is sometimes called an "identical rhyme" by pop songwriters.) Homonyms are considered identities, although they are spelled differently and mean different things, because they *sound* iden-tical. The words *two, to,* and *too* in a lyric are all identities. Identities may be used occasionally for a specific effect, but they are not substitutes for perfect rhymes. In theater songs, identities sound like inept or unimaginative attempts at rhyme.

It's important to understand the difference between rhyme and identity, so here are a few more examples. *Astronomical/anatomical* is a rhyme, but *astronomical/economical* is an identity. If this seems confusing, remember that a rhyme only involves the last accented syllable and any unaccented syllables that follow. So with *astronomical/anatomical* the relevant syllables (*nom*-ic-al and *tom*-ic-al) rhyme, because the initial consonants are different while everything that follows is identical. But with *astronomical/economical* the syllables (*nom*-ic-al and *nom*-ic-al) are exactly the same, *including* the initial consonant—making an identity, not a rhyme. Similarly, *graduation/gradation* is a rhyme (*a*-tion and *da*-tion), but *graduation/radiation* is considered an identity: When these words are sung, what we hear in the relevant syllables is identical, namely *a*-tion and *a*-tion. (If you doubt this, try singing "grad-u-*a*-tion" and "ra-di-*a*-tion" slowly.)

So-called *near-rhymes* may match vowels but not consonants (like *line/find*), or match consonants but not vowels (like *past/rest*). Near-rhymes are accepted in popular music, but in musical theater they are considered a sign of sloppy or inferior craftsmanship, and are avoided by good songwriters. Among the most common types of near-rhymes are singular with plural nouns (*day/ways*); present with past tense (*retire/fired*); and words ending in *m* with words ending in *n* (*time/mine*).

Something else that good songwriters watch out for is the inadvertent use of near-rhyme. It is natural and rather common for near-rhymes to occur in places where they will seem to be attempts at rhyme. Here is an example.

> Things were different then,
> So many things went wrong;
> Now you are my only friend
> And now our ties are strong.

Even if the first and third lines don't rhyme in other stanzas, it still sounds as though *then* and *friend* were intended to rhyme in this one. Because such near-rhymes are unintentional, it can be difficult for the lyricist to notice them. For this and other reasons, many lyricists put a lyric aside and then come back to it after a day or so. At that point it is possible to be more objective about their work and to see flaws, like near-rhymes that may have been written unintentionally. A lyricist who is still unsure may want to let a collaborator look at the lyric; a second pair of eyes may find things that even the lyricist's own close examination has missed.

Sometimes lyricists will deliberately twist the pronunciation of a word in order to make it rhyme. E. Y. "Yip" Harburg was a master of this technique. In *The Wizard of Oz*, the Cowardly Lion's reprise of "If I Only Had a Brain" includes such rhymes as *prowess/mowess* and *no denyin'/dandelion*. This is an effective technique, but it is best when used sparingly, and only where appropriate. It works here, because it comes from a film aimed at children; because the Lion, as played by Bert Lahr, has already been established as a comic character who uses colloquial elisions like *denyin'*; because the mood of the scene has been one of whimsy and humor; and finally, because the rhymes call attention to precisely the words that are most important—and funniest—in both stanzas.

In context, both alterations come across as charming and funny. But if Harburg had made such an alteration in a serious song like "Over the Rainbow," the effect would have been inappropriate and strained.

Similarly, *Kiss Me, Kate*'s hoodlums, singing "Brush Up Your Shakespeare," get away with rhymes like *flatter 'er/Cleopaterer* and *British Embessida/Troilus and Cressida*, because they are essentially comic caricatures in an out-and-out vaudeville number full of outrageous puns; it would be harder to justify such rhymes or puns during a ballad like "So in Love." In *Company*, Bobby's three girlfriends are not presented as caricatures, but in "You Could Drive a Person Crazy" they get away with such rhymes as *personable/coercin' a bull* because from the very beginning the song is set up as a wacky presentational number.

Good lyricists keep in mind that lines that are set up to rhyme *must* rhyme, which means a true rhyme—even with twisted pronunciation—rather than a near-rhyme. If you imagine the Cowardly Lion singing "But I could show my prowess, /Be a lion, not a mou-ouse," you can hear how lame the near-rhyme sounds compared to the true rhyme.

In theater lyrics, moreover, a rhyme for one character may not be a rhyme for another. For example, British people with upper-class accents have traditionally pronounced "been" to rhyme with "queen." This is not true for Americans, who rhyme "been" with "tin." Among John Latouche's generally excellent lyrics for *The Golden Apple* is an egregious violation of this principle, when he rhymes "this you" with "issue." The characters singing, people from a small town in Washington state, might rhyme "issue" with "fish you," but not "this you."

This is not to say that all lines, even prominent ones, must rhyme. *Not* to rhyme is sometimes even more effective, as the chorus of "Ol' Man River," from Oscar Hammerstein and Jerome Kern's *Show Boat*, demonstrates. In the first stanza of the chorus, rhyme is absent. Instead, the end of the second line repeats the first line, and each of the next three lines has a different "feminine" ending (*sumpin'/nuthin'/rollin'*), to which the last line adds one word, "along." (A "feminine" ending or rhyme in poetic terminology means an accented syllable followed by one or more unaccented syllables, while "masculine" lines end on a single accented syllable.) The second stanza maintains this incessant rhythm, perhaps suggesting the endless flow of the river, but adds a rhyme, *cotton/forgotten*. About this lyric, Hammerstein later commented that to rhyme where it is unnecessary can focus the listener's attention on the rhymes, rather than on the song's meaning. "If, on the other hand, you keep him waiting for a rhyme, he is more likely to listen to the meaning of the words."

Once Hammerstein introduces a rhyme, however, he does not go back. The B stanza has two rhymes (*strain/pain* and *bale/jail*), and the last A stanza has a rhyme in the same place as the second A (*tryin'/dyin'*). To go back to nonrhyme after introducing rhyme would seem like a regression, as though the lyricist were unable to find a rhyme. Even if deliberate, it would still sound like a mistake or an omission.

It is even possible to write a lyric without a rhyme or regular metric pattern, although this is more difficult than it seems. One example is the song "Frank Mills" from *Hair*. The classic ballad "I Talk to the Trees" from Lerner and Loewe's *Paint Your Wagon* goes almost as far; there is only one rhyme in the entire chorus.

Rhymes appear most often at the end of lyric lines, where they are sometimes called *end rhymes*. Lyricists also use rhymes in the middle of lines, where they are called *internal rhymes*. Stephen Sondheim has said that internal rhymes can add punch or drive to a lyric. They can also make it flow faster, or slow it down. Here are a few examples of internal rhymes.

This is the first A section of Irving Berlin's "Everybody Step":

Everybody step to the syncopated *rhythm*,
Let's be goin' *with 'em* when they begin.
You'll be sayin', "*Yessir*, the *band* is *grand*,
He's the best pro*fessor* in all the land."

This is the bridge of Berlin's "A Pretty Girl Is Like a Melody":

Just like the *strain* of a haunting re*frain*,
She'll start u*pon* a mara*thon*
And run around your brain.

And this is the bridge of Stephen Sondheim and Jule Styne's "Small World":

We have so much in *common*,
It's a phe*nomen*on.
We could pool our re*sources*
By joining *forces*
From now on.

Titles may be rhymed or unrhymed. In "Comedy Tonight," from *A Funny Thing Happened on the Way to the Forum*, as in many other songs, the title is not rhymed until the last stanza of the chorus.

There are many rhyming patterns available to lyricists. They can, for example, rhyme every pair of lines (*aabb*). Another possibility is to rhyme every other line (*abab*). There are so many possibilities, in fact, that aspiring lyricists would do well to look at the libretto or vocal score of any good show and examine the schemes used. Except in extraordinary circumstances, lyricists try to keep the number of songs with the same rhyme scheme to a minimum. Boredom may set in once listeners detect a pattern. Good lyricists, by varying end-rhyme schemes within a show and using internal rhymes from time to time, keep their lyrics fresh and interesting.

One more word about surprise and predictability in rhymes: Ordinarily, a writer doesn't want the audience to be ahead of him. If the audience can see an obvious rhyme approaching, it will be bored or annoyed by the time the song gets there. On the other hand, there are a few comedy songs in which part of the fun is knowing a particular rhyme is coming and enjoying it in advance. This only works in a tremendously funny song, however—a song that the audience already loves so much that it can enjoy the rhyme in anticipation.

Three Principles

CONCISION

More than any play, any short story, and most poems, a song lyric must be succinct. Every word counts. This means that lyricists can rarely afford to waste time on an empty word or phrase such as "well," "quite," "just," or "you know." It also means that every little word—even "a" and "the"—needs to be absolutely right. Stephen Sondheim cites the first line of DuBose Heyward's lyric for "Summertime" in Act One of *Porgy and Bess* as an example. Sondheim says that another lyricist would have written "Summertime *when* the livin' is easy," which would have been boring compared to Heyward's "*and*"; in Sondheim's opinion, *and* is the precisely right word, "and that word is worth its weight in gold."

COMPREHENSIBILITY

One of the most important things for any lyric to be is comprehensible. Most lyrics are heard only once. With instruments playing underneath, and the sonorous, nonsemantic effect of rhyme—not to mention the other distractions of scenery, lights, costumes, and dancing—lyrics are often harder to hear and understand than ordinary dialogue. In addition, because lyrics are shorter and presumably more concise than dialogue, it is even more essential that every word of a lyric be understood. Frank Loesser once likened a song to a quickly moving train, which passes by and then is gone. To a large extent the speed is under a lyricist's control. For instance, while clever twists and multiple rhymes can show off a lyricist's skill, too many twists or rhymes in a lyric make it harder to understand.

Comprehensibility also usually requires that a lyricist eschew such "poetic" practices as inverting word order. Good lyrics rarely invert idiomatic word order. "I've got the world at my fingertips" sounds far more natural than "At my fingertips I've got the world." Good songwriters also avoid archaic language, unless it is both comprehensible and necessary to give a flavor of the time and place. The exception, as with many such principles, lies with comedy songs. At the end of "Brush Up Your Shakespeare," the gangsters toss in "Odds bodkins!" with no regard for its actual meaning, which only adds to the humorous incongruities of the song.

For shows set in another time and place, lyricists follow the same principle as librettists: They use actual archaisms for flavor only, because a little goes a long way. Sherman Edwards's lyrics for *1776* are generally in contemporary English, but they are sprinkled with real (or real-sounding) archaic usage, like "God, Sir, get thee to it," which gives a flavor of the late eighteenth century. Similarly, while there are no ethnic words in *Fiddler* besides the Russian words and the three Hebrew phrases previously mentioned, Sheldon Harnick's lyrics contain some typical examples of Jewish immigrant usage (in English, of course), such as "A blessing on your head," that help Joseph Stein's book to establish the atmosphere.

But comprehensibility is not simply a matter of clarity of thought and language; it also depends on the lyrics and music matching each other. Unlike a poem, which can be read over and over, a lyric is heard in time and to music. No matter how clear and simple the lyric is, it will not be comprehensible unless the music works with it and not against it. Take as an example the first line of the chorus of "People Will Say We're in Love." Rodgers set Hammerstein's lyric "Don't throw bouquets at me" so that there is a metric accent on "don't" and a stronger accent on "[bou-]quets." Now imagine instead that the first accent is on "throw" and the strong accent on "at." Already the lyric becomes harder to understand. Or imagine that the final "me" is set, not to a low note, but to the highest note in the line, so that it sounds like a question rather than a statement. Again, we may still understand the lyric, but it has become harder.

Of course, if audience members ponder a badly set line and its context, its meaning will usually become clear. But while they are pondering, they are likely to miss the next several lines and—assuming the lyric is concise—a small but significant part of the story as well.

The comprehensibility of a lyric also depends on how *singable* it is. There are many potential stumbling blocks to singability. It isn't possible to catalogue them all—songwriters need to rely on their own taste, common sense, and experience—but here are some examples.

When one word ends with a particular consonant and the next word begins with the same sound or a related one, it is often hard for the singer to make a distinction between them, and hard for the listener to hear that distinction. For instance, if the line is "It's a time of festival," some listeners may hear it correctly, while some may hear the two *f*'s as a single one—for instance, as "It's a time, a festival." Or if the line is "A local lady gave me help," some may hear it as "A local aide, he gave me help."

Some combinations of sounds are inherently difficult for a singer to enunciate clearly. A line containing a phrase like "mismatched charms," with its densely packed consonants, is awkward for a singer and confusing for the audience. Sibilants like *s* and *z* can clog up a line. (Alan Jay Lerner said that after completing a lyric, he would go over it and try to remove every sibilant he could.)

Words that end in consonants, especially unvoiced strong consonants like *t*, *k*, and *p*, can be difficult to sing and sustain, and often unpleasant to hear. Vowel sounds in certain ranges can also be a problem. Many singers, for example, have problems singing high notes on closed vowels like *ee* or *oo*. ("Closed" and "open" refer to the position of the mouth and jaw.)

PARTICULARIZATION

Good lyrics are specific to the show for which they are written, growing organically out of the story and the characters. They are also specific to the character who sings them, reflecting that character's vocabulary, grammar, colloquialisms, elisions, pronunciation, and current state of mind.

The lyrics of "Wouldn't It Be Loverly?" reveal that beneath Eliza Doolittle's rough exterior lie warmth and sensitivity. Notice that the song is not about any of these aspects of Liza's character, but about her wish for comfort and love. Liza doesn't tell us "I'm ignorant, I'm sensitive, I'm warm"; these things are revealed, in passing, by the way she tells us what she wants. The *way* in which a character talks or sings can tell us as much about her as *what* she talks or sings about.

Particularization is essential for two other reasons. First, if founded on what the character thinks and the way the character talks, particularization can give the lyricist ideas. For instance, in the song "Now" from *A Little Night Music*, Frederick Egerman contemplates trying to entice his virginal young wife into having sex. He is a successful lawyer, so it is plausible that he would plan things methodically, as he does in this song: Every course of action is carefully considered, and subdivided into a plan A and a plan B. When he imagines reading aloud to get his wife in the mood, as an educated man at the turn of the twentieth century his thoughts turn to de Sade, Dickens, and Stendhal. On the other hand, when the Broadway gamblers of *Guys and Dolls* discuss the likelihood of an occurrence, they speak of it as "better than even money" or "a probable twelve to seven," and when Sky Masterson talks to Sarah about his previous understanding of love, he says he thought that he "knew the score."

A second reason for particularization is that it helps to focus the audience's attention. For example, if you watch a television news story that simply tells you that a big corporation has laid off ten thousand people, you may think "Tsk tsk," but probably not much more. If instead the news story focuses on one of the longtime employees who has been let go, a single mother with three kids who doesn't know how she will buy food or pay the rent next week—and if it shows you her children happily playing, unaware that they are about to lose everything—you're much more likely to care and to remember. It's individual details, not vague ideas, that usually evoke emotion.

Must lyrics be specific to the character and to the show? If a song is needed in a certain spot, would it be so bad to take something written as a pop song, or for a different show, and put it into this one? The short answer to both questions is *yes*. It's true that before the "golden age," songs would often be inserted into shows for which they were not intended. Indeed, it was not uncommon in the 1920s for a songwriter to have songs inserted into shows that had scores by other writers. But that was a very different time, when almost no songs were specific to shows and to characters, nor was there any significant distinction between theater songs and popular songs. Insertions were not as painfully obvious as they would be in one of the classic "golden age" shows.

This is not to say that songs have never been inserted into shows since 1943. But such insertions have usually been made in revues or in light, patchwork book shows, and rarely in good, well-integrated shows.

Imagine, for example, that to show Laurey and Curly's relationship at the beginning of *Oklahoma!* Rodgers and Hammerstein inserted "If I Loved You" (from *Carousel*), which they happened to have lying around, instead of writing "People Will Say We're in Love." The basic situation for each song is not very different: A boy and girl who are attracted to each other don't want to admit it. Would "If I Loved You" work? Perhaps, but not very well. For one thing, it is a serious ballad, solemn and even melancholy, with intimations of tragedy ahead. Curly and Laurey are much happier, healthier characters than Billy Bigelow and Julie Jordan; their relationship is lighter and sunnier, with plenty of banter and teasing. Though "People Will Say We're in Love" is a ballad, it maintains a bantering tone throughout, and the depth of each character's feelings for the other emerges only at the very end of each chorus. In addition, *Oklahoma!* is a much more lighthearted and comedic show than *Carousel*; while we can imagine Curly being killed, it would be hard to imagine him coming back after death to look after Laurey. The lyrics of "If I Loved You," and the music as well, would feel false and wrong in *Oklahoma!*

The better the show—the tighter the integration among the book and the lyrics, the plot and the characterizations—the more such differences of tone will jar the audience. Even if only on a subliminal level, such discontinuities and incongruities can tear the delicate fabric of the theatrical illusion of organic life that the writers are attempting to create—even in farces like *A Funny Thing Happened on the Way to the Forum* and stylized shows like *Grease*. When a song is well integrated with the show, as Lehman Engel says, we hear "not merely a song but a piece of theater—a segment of an entire continuing fabric."

The Balance of Illusions: Poetry versus Conversation

Lyrics in musical theater songs are presumed to be "speech." To be precise, they are heightened speech. A good lyric sounds natural and conversational. The language is consistent, and does not suddenly lapse into inappropriate slang, dialect, or vocabulary. Lyrics also resemble poetry in that they rhyme and maintain regular metric patterns—and yet, as we have seen, lyrics are not quite poetry either. Rather, good lyrics balance the illusions of conversation and of poetry.

One way lyricists do this is by the use of poetical devices such as metaphor, simile, hyperbole, personification, and onomatopoeia, many of which we also use in everyday speech. A *metaphor* compares one thing or idea with another by equating them. When we say of a slovenly man that he is "a pig," we don't mean that he is literally a swine, of course, but that his behavior seems piglike. When Reno Sweeney sings "You're the Colosseum" in *Anything Goes*, she isn't talking to an ancient edifice in Rome, and when Sally Bowles sings that life is a cabaret, she doesn't literally believe that life consists of a large piano bar full of people singing show tunes.

A *simile* also compares one thing or idea to another, but uses "like" or "as," as when we say that someone is "as strong as an ox." In Rodgers and Hammerstein's "It Might as Well Be Spring" from *State Fair*, a young girl sings that she is as jumpy as a puppet on a string.

We often use *hyperbole*, or exaggeration, in our speech. "I'm so hungry I could eat a horse" is hyperbole for any normal human being. Hyperbole is common in musical theater lyrics. When in *West Side Story* Tony sings that Maria is the most beautiful sound he has ever heard, or Maria tells Tony that he is the only thing she will see, hyperbole is a basic part of the lovers' vocabulary.

Personification gives to things the attributes of human beings. For example, sailors traditionally have referred to ships as "she." The entire lyric of "Luck Be a Lady Tonight" is an example of personification.

Personification of parts of the body, especially the heart, has been overused and abused in song lyrics. Alan Jay Lerner wrote of his song "I Could Have Danced All Night" that he was always embarrassed about the line "Why all at once my heart took flight." He said he had always loathed lyrics in which the heart skips or leaps or jumps. He meant to change the line but never did. We can all understand the meaning of a simple metaphorical opposition such as "My head says this but my heart says this," but it is hard to imagine anyone saying that her heart wants to sigh like a chime that flies on a breeze, as Maria does in *The Sound of Music*.

Onomatopoeia consists of words that approximate the sounds they describe. We say that bees *buzz* or that an old clock or watch makes a *tick-tock* sound. In Meredith Willson's "Seventy-Six Trombones" from *The Music Man*, Harold Hill sings that as a tuba player in a marching band he once "oom-pahed" throughout the town.

Other poetic devices, like alliteration, repetition, and parallelism, are not found so often in everyday language, but lyricists may find them useful. *Alliteration* is the repetition of initial consonants in two or more words, as in the old tongue twister "Peter Piper picked a peck of pickled peppers." When used prudently, it can add rhythmic drive, or a sense of fun, to a lyric. The lyric for "It Might as Well Be Spring," mentioned earlier for an example of simile, also makes good use of alliteration, such as "a willow in a windstorm."

Repetition of words or short phrases is rarely used intentionally in everyday language, except when our emotions are high or we attempt to control events by saying "Stop, stop!" or "No, no, no!" But for both expressive and formal reasons, repetition is often found in poetry—think, for instance, of Walt Whitman's "O heart! heart! heart!" or Shakespeare's "Never, never, never, never, never!" from *King Lear*. Repetition is similarly common in lyrics.

Parallelism of thoughts or phrases, at least in written prose, is more common than simple repetition; it often occurs in expressions like "the more things change, the more they stay the same" or Caesar's "I came, I saw, I conquered." And parallel lyrics, which work so well with parallel musical phrases, can be found in virtually every good song ever written.

There are, however, some poetical devices that do not translate well into lyrics, including consonance and assonance. In *consonance*, two words that have similar consonants but different vowels are linked together, such as *song* and *ring* or *bland* and *blind*. While this works in certain styles of poetry, too often in lyrics it simply sounds like near-rhyme, or bad rhyme. There can be exceptions, however, when the lyricist makes clear that consonance and not near-rhyme is intended, as in the first chorus of Jones and Schmidt's "Try to Remember" from *The Fantasticks*.

Try to remember the kind of September
When life was slow, and oh, so mellow.
Try to remember the kind of September
When grass was green and grain was yellow.

Try to remember the kind of September
When you were a tender and callow fellow.
Try to remember, and if you remember,
Then follow.

Here a feminine rhyme is set up at the end of the second, fourth, and sixth lines (*mellow/yellow/fellow*), but the sixth line also utilizes consonance (*callow fellow*), which was already introduced internally in the fourth line (*green* and *grain*). Since the eighth line ("And follow") is drastically shorter than the other even-numbered lines, the use of consonance rather than rhyme at the end of this line, while perhaps a little surprising at first, is clearly intentional. (Many writers feel that *September/tender* is perilously close to a near-rhyme.)

Assonance is the repetition of vowel sounds throughout a line or stanza. A couplet like "I never felt better/In wet fresh weather," with all its short *e* sounds, is an extreme example of assonance. Assonance usually sounds like near-rhyme at the ends of lines, where good lyric writers usually avoid it. Nevertheless, lyricists occasionally make use of it in the middle of lines to strengthen the overall effect of a stanza, as Tom Jones often does.

It is this balance of everyday language with the language and devices of poetry that makes ballads, in some ways, the most difficult lyrics to write. If a lyric reads like poetry, especially pre-twentieth century poetry, it is probably too flowery and artificial to make a good lyric. On the other hand, ballad lyrics, if read without the music, may seem flat or dull. But they are written to be set to music, which helps to float them above the level of prose. For example, Stephen Sondheim's lyrics for "Send in the Clowns," from *A Little Night Music*, seem quite conversational. But this lyric has a poetic quality beyond the rhyme scheme. The ingenious choice of a title line, with its connotations of the circus and of farce, is one example. Another is the use of theatrical imagery (such as "making my entrance" and "losing my timing"), which is appropriate both to the "clowns" metaphor and to Desiree's career as an actress. Yet another is the way Sondheim delays the rhyme of the third line ("ground") until the third line of the second stanza ("around"), where it sounds unexpected but nevertheless natural. By balancing poetry and conversation, good songwriters can create this balance of unpredictability and inevitability.

Sincerity helps to keep the poetic and conversational aspects of a lyric in balance. Good lyricists write for characters as if they were the characters, and truly believe what the characters are saying. Oscar Hammerstein II advises lyricists to say what they mean from the bottom of their hearts, and say it as carefully, clearly, and beautifully as possible.

CHAPTER 6

THE MUSIC

Song Basics

MELODY

In the abstract, melody is a succession of single pitches and silences forming a unified line in time. Melodies, like sentences, can be divided into smaller units or *phrases*, which themselves can be broken down into their component subphrases. Subphrases can often be broken down into still smaller thematic units, all the way down to motifs. A *motif*, or motive, can be described as a short thematic kernel. Usually it consists of a few notes in a particular rhythm. Let's look at the motivic construction of "Frère Jacques."

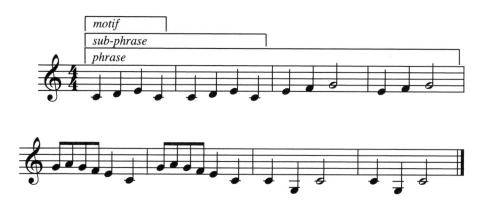

Motif #1 coincides with the words "Frère Jacques," and it is immediately repeated to make up the first subphrase. The phrase is completed by once again using and repeating the motif, but this time starting a third higher and eliminating the falling note. The next phrase begins with a different motif—moving in the opposite direction to the first motif—that is also repeated immediately to make up a subphrase.

A distinctive motif helps to make a song more memorable. But what makes a motif "distinctive"? Many composers create motifs based on particular rising or falling melodic intervals, in combination with interesting rhythms. One of the most famous motifs is found at the beginning of Beethoven's Symphony no. 5 in C Minor. The motif consists of a note played three times followed by a downward leap of a third, coupled with one of the most catchy and recognizable rhythms ever created.

A particular significant interval is a characteristic element of many motifs. Here are a few examples of motifs from popular and theater songs, with their significant intervals marked.

Once composers have a motif, what can they do with it? They can repeat it, of course, as the first three notes of " 'Swonderful," with their falling third, are repeated. Or they can invert it (turn it upside down), as in the first two A sections of "People Will Say We're in Love." In these sections, each of the first three lines begins with a rising perfect fifth and ends with a falling fifth.

Another way to exploit a motif is to repeat it on different pitches, making a *sequence*. Classical composers use sequences all the time. Let's take another look at Beethoven's Fifth Symphony, first movement, for a clear-cut example.

"Everything's Alright" from *Jesus Christ Superstar* also uses a sequence. Each of the first three bars of the melody is a one-measure subphrase, and each begins one note lower than the previous one.

There are other ways to use a motif. One way is to vary it by gradually making the characteristic interval wider, as in "The Surrey with the Fringe on Top" from *Oklahoma!* The first two-measure subphrase of the melody ends with a leap up of a perfect fourth (on "scurry"). The next two-measure subphrase begins identically, but ends with a wider leap, up a perfect fifth (on "surrey"). The next phrase leaps up a major sixth (on the next "surrey") before continuing on for another two measures.

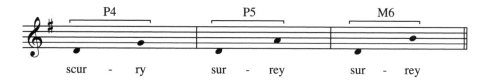

The opposite technique is to take a motif with a large interval and gradually contract it. A clear example is in "Bewitched" from *Pal Joey*, in which a three-note motif begins with a span of a minor sixth (on "wild again") and gradually contracts to a simple two-note motion of a minor second (on "bewitched").

Motifs can also be varied by adding "nonharmonic tones." A neighbor tone, for instance, moves a step above or below the previous pitch and then immediately returns to it. Passing tones fill in the musical gap between two notes that are a second or a third apart. An *appoggiatura*, which literally means a "leaning note," is a nonharmonic tone on a strong beat that resolves either upward or downward by step to a harmonic tone.

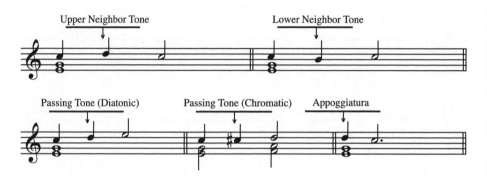

Other techniques include augmentation (lengthening the duration of one or more of the notes) and diminution (shortening their duration). The mode can also change from major to minor, or from minor to major. All of the above ideas are useful in building motifs into themes, and phrases into full songs.

A melody may stay on one pitch for a while, but eventually it moves from one note to the next by step, by leap, or by a mixture of the two. Most melodies move by step more often than by leap.

Large leaps are often followed by stepwise motion in the opposite direction, essentially "filling in" the gap between the low note and the high note of the leap. An example of this can be found in the song "Sunday" from *Sunday in the Park with George*, in which the initial upward leap of a minor sixth is followed by primarily stepwise motion that gradually fills in the notes between.

Collectively, the various pitches used to create a melody make up its *melodic range*, that is, the total musical span from the lowest note to the highest one. A song's melodic range can be small, as in songs written for young children; wide, as in many operatic arias; or somewhere in between, as in most popular music. For instance, Gregorian chant tends to have a very small range.

Sometimes the range can be extremely wide, as in the case of "Pity the Child" from *Chess*. The melody extends from a low B-flat to a D-flat more than two octaves higher.

Most melodic shapes are based on an arc, either rising and then falling (a sort of hill shape) or falling and then rising (a valley shape). The hill-shaped arcs are the most prevalent among melodies, with valley-shaped arcs a close second. You may have noticed a similarity between the rising arc of a melody and the dramatic arc discussed in chapter 3. We don't believe this is a coincidence; it seems likely that behind both arcs is a common psychological pattern that people find satisfying.

Rising ("hill") arc   Falling ("valley") arc

Melodic arcs often return to, or near, their starting pitches. "Frère Jacques," with its hill-shaped arc, begins and ends on the same note; so does the chorus of the title song from *Oklahoma!* (a valley shape). Many melodies, arc-shaped or not, move toward an important musical goal at some point, usually about two-thirds or three-fourths of the way through. We call this the *climax* of the melody. The climactic note is usually the highest pitch of the entire melody, although sometimes it is the lowest; it often coincides with the lyric's most important word.

"My Funny Valentine," from *Babes in Arms*, is a case in point. It starts low, and the range in the first four bars is narrow. That changes almost immediately with the words "You make me smile," with a leap up to a note a minor seventh above the tonic on the word "smile." There is another leap up on the word "fav'rite," this time taking the melody up a ninth from the tonic. After dropping back down a bit, the melodic range again expands. It finally reaches the highest note—a full octave and a minor third above the melody's lowest note, the tonic—on the most important word of the song, "stay."

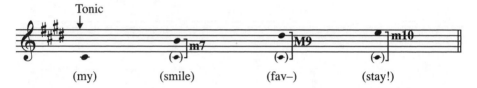

Melodies are *diatonic* to the extent that they stay exclusively within the key of the song and use no accidentals (sharps, flats, or naturals). Melodies that range into other keys or use many accidentals are *chromatic*. Folk songs and children's songs, like "Frère Jacques," tend to have diatonic melodies. Jerry Herman uses diatonicism to good effect in "Before the Parade Passes By" from *Hello, Dolly!* and in "I Will Follow You" from *Milk and Honey*. Other musical theater songs with diatonic, or mostly diatonic, melodies include "Happiness" from *You're a Good Man, Charlie Brown*, "Try to Remember" from *The Fantasticks*, "What Makes Me Love Him" from *The Apple Tree*, and "You'll Never Get Away from Me" from *Gypsy*.

Harmonies, like melodies, can be diatonic, chromatic, or something in between. Sometimes a chromatic melody can enliven a diatonic accompaniment, as in Irving Berlin's "Everybody's Doing It Now" from *Watch Your Step*. Here is the beginning of the actual song, and then a diatonic version, with the chromaticism eliminated. As you can see, the chromatic, nonharmonic tones keep the melody from being bland; they give it individuality and interest.

Original version

Diatonic version

There are few rules as to when diatonicism or chromaticism is more appropriate in a song. Relatively speaking, diatonicism sounds uncomplicated and straightforward. In major keys, it is effective in hopeful or happy situations. In minor keys diatonicism can be equally clear-cut, although songs in minor usually contain some chromaticism for a variety of reasons. Chromatic music can imply many kinds of emotional states. For instance, it can sound sexy, devious, or melancholy.

Setting Lyrics

Music unfolds over time. Rhythm is often considered the most fundamental aspect of music, because it defines how time is partitioned by sound and silence. While you do not need melody or harmony to make music, you cannot make music without rhythm. Rap, which began in the late 1970s and was still one of the most popular styles at the beginning of the twenty-first century, illustrates how fundamental rhythm is to music. Much rap has no singing at all.

When we speak of rhythm in musical theater, we are really dealing with three different things: the rhythm of words, the rhythm of melody, and the rhythm of accompaniment. For the moment, let's concentrate on the rhythms of words and melody.

Whichever is written first, a good melody matches the natural meter of the lyrics and the natural accents of the words. The general principle that composers follow in setting musical theater lyrics is one note per syllable; melisma (more than one note per syllable) is limited and infrequent, because it tends to obscure intelligibility and meaning. An example of this is the title song from *Mame*, in which only the title name, at the end of the first and second lines of each A section, is given a short melisma while the rest of the lines are syllabic.

Good songwriters strive to make their songs both singable and understandable by making accented syllables coincide with strong beats or the stronger parts of a beat, and by setting words consistently. In setting lyrics from a rhythmic viewpoint, a composer must examine the prosody of the lyrics—the natural rhythms of the words, including their stresses and intonation. Otherwise, he is in danger of "ac-CENT-ing the wrong syl-LA-ble" or emphasizing an unimportant word. A diligent composer reads through a lyric several times to find the naturally weak and strong syllables. He tries it with different emphases, listening carefully for which rhythms sound better and which don't work at all. If he's at all unsure about the correct pronunciation or accentuation of a word, he refers to a dictionary.

Distortions of natural speech and meter are sometimes found in opera and more often in pop music, but they are unacceptable in musical theater, where comprehension and clarity of meaning are paramount. Accenting syllables that are naturally unemphasized, or stressing unimportant words like "the" or "of," confuses listeners. Even experienced writers can set words inappropriately; "Razzle-Dazzle" from *Chicago* emphasizes a number of insignificant words, as in "stiffer *than* a girder." Such distortions represent stumbling blocks that singers must overcome in order to give an effective performance.

Setting a lyric is not just a matter of setting words with natural rhythms; it's also making sure that those rhythms, and the notes, make sense both musically and dramatically. Every lyric line contains a word or two that is more important than the rest. A composer must know not only which words need to be emphasized to match the metrical rhythms, but also which words need to be emphasized to produce the intended meaning. To illustrate this, let's examine a sample lyric line for its potential rhythms.

The world is a lovely place.

There are a number of possibilities for setting this lyric rhythmically. We could make a case for "world," "is," or "lovely" as the most important word. A character could say "The *world* is a lovely place," using the natural metrical accent on "world," if she feels that the universe is benevolent, or nature is beautiful, or she is simply feeling happy. Or she could say "The world *is* a lovely place" to disagree with someone who says that the world is not a lovely place. (This is the only possibility that is not suggested by the natural metrical rhythm.) Alternatively, she could say "The world is a *lovely* place" if she sees a breathtaking landscape for the first time or from a new perspective. But while we might assume that "place" could also be emphasized, upon examination we can see that it doesn't warrant the same stress that the other words might. Saying "The world is a lovely *place*" does not make much sense. (A place as opposed to what?)

One way to stress an important word is to place it on a strong beat or strong part of a beat. Other techniques are to make the word noticeably higher or lower than the rest of the melodic phrase, or to sustain it longer.

To highlight a lyric, composers sometimes use "word painting," imitating in the melody the concept presented by the lyrics. A phrase like "reach for the sky," for example, may have an ascending melody, or a descending melody could be used for "my heart is sinking." In "On the Street Where You Live" from *My Fair Lady*, the chorus begins with three notes moving stepwise before a leap up to a sixth above the tonic. The next line (beginning "But the pavement always stayed") suggests that the character is not as grounded as he had been. The musical phrase begins like the first one, but now it jumps to the seventh above the tonic before pushing on. The third phrase (beginning "All at once") begins the motif one step higher; then it makes a large leap up to a tenth above the tonic, which adds to the feeling of soaring.

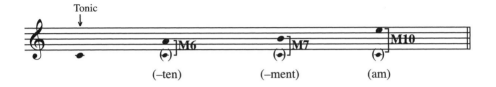

In the previous chapter we talked about a song's lyrical structure, and how it can be analyzed in terms of sections such as verse, chorus, and so on. The musical structure ordinarily matches the lyrical structure, often as a 32-bar form that in turn can be broken down into eight-measure phrases. The cadence points at the end of each phrase match the primary rhymes, and harmonic shifts usually reflect lyrical shifts in perspective.

Once in a great while the structure of the music doesn't exactly match that of the lyric. As mentioned in the previous chapter, the lyrics of "You'll Never Walk Alone" use an AABC pattern instead of the usual AABA form. The music goes a step further: It is in four related but different sections, ABCD. On the other hand, in "Frank Mills" an irregular lyrical form is set to a regular AABA musical pattern.

ACCOMPANIMENTS

Singing doesn't always need accompaniment. Unaccompanied, or *a cappella*, singing is the norm for church chants (as in the opening of *The Sound of Music*) and barbershop quartets (as in "Sincere" from *The Music Man*). Folk songs are often sung unaccompanied by soloists and choruses, and the title song of *Brigadoon* is sung *a cappella* at the beginning of the show as though it were an authentic folk song.

Unaccompanied singing isn't limited to these types of songs, however. A solitary voice singing a short, plaintive song or phrase without instrumental accompaniment can sometimes be more emotionally powerful than the same song backed by a pit orchestra would be. Still, *a cappella* singing in musical theater is infrequent, unless a show's premise requires it, as a "doo-wop" musical might.

Accompaniments, like theater songs in general, reflect the times in which they were written. In the first half of the twentieth century, the most common style of accompaniment used the alternation of a bass note with a chord. It was based on the popular dance rhythms of the time, particularly the fox-trot. Veteran songwriters often refer to this style of accompaniment as "boom-chick."

A similar accompaniment in triple meter is the "oom-pah-pah" used for waltzes.

Since the mid-1950s the dominant trend in popular music has been rock in all its varieties. The chief rhythmic feel of rock has been the 4/4 backbeat, with accents heard or implied on the weak beats, two and four. The meter is the same as that of the "boom-chick" accompaniment—the notes may even look similar—but the rhythmic effect is very different.

Because of the broad range of rock styles, the backbeat is nowhere near as universal as the "boom-chick" rhythm was before, but rock has made an indelible mark on composers' musical thinking. In addition to musicals by rock composers, such as *The Who's Tommy* by Pete Townshend, *Aida* by Elton John, and *Chess* by Benny

Andersson and Björn Ulvaeus, more and more musicals in the last few decades have been heavily influenced by rock, especially in their accompaniment styles. Since the 1970s, fox-trot accompaniments have been increasingly rare. Unless they are used deliberately for comedic or pastiche effect, they sound tired and dated.

Rock's influence extends to accompaniment figuration. Repeated chords reminiscent of doo-wop and other early rock 'n' roll styles can be found in "There Are Worse Things I Could Do" from *Grease* and "Grow for Me" from *Little Shop of Horrors*. Other accompaniments use arpeggiation based on guitar techniques. In *Company*, for instance, "Someone Is Waiting" uses a pattern of a bass downbeat with rolled chords on the second beat typical of some folk guitar styles, while "Another Hundred People" is reminiscent of finger-picking, in which the fingers pluck different guitar strings to alternate low and high notes.

Bass lines have also been affected by rock. Songs of musical theater's golden age relied heavily on unsyncopated bass notes, often in a "walking" style. Rock-influenced theater songs, by contrast, use bass lines that alternate long notes and shorter ones.

Typical "Golden Age" bass line   Typical Rock-influenced bass line

Harmony has been affected by rock as well, although more subtly. In some shows from the 1940s and 1950s, the influence of jazz was heard in the use of chords with added tones, such as a major chord with an added sixth and ninth; extended chords, such as minor ninth or eleventh chords; and extended chords with altered tones, such as a dominant seventh with an added flat ninth.

Another harmonization technique common in jazz of the 1940s and 1950s was chord substitution. Substitutions often rely on tritone relationships between an expected chord and its replacement. A composer might replace the dominant chord in a ii-V^7-I progression, for example, with a flatted II or Neapolitan chord (for instance, D-flat seventh, instead of G seventh, going to C major). Chords can be altered or extended for a "jazzier" flavor. Many songs from *City of Angels* use both extended chords and substitution. However, consistent use of such techniques sounds dated, and it is mostly suitable for pastiche, as in *City of Angels*.

Rock music has also had an influence on musical theater harmony. One notable example is the weakening of a previously almost universal dependence on functional harmonic progressions. (Harmony is "functional" when it relies on motion through the circle of fifths to end on a V-I cadence.) This change can be seen most clearly if we compare cadences in songs written before the rock era to those written since.

Virtually every cadence in songs from shows written before the late 1960s uses functional progressions, and virtually every song ends with a perfect cadence (a V-I or V^7-I cadence in which the melody ends on the tonic note). Here are typical examples from the 1930s, 1940s, and 1950s.

Many rock songs, however, avoid perfect cadences, and a good number avoid functional progressions as well. Since 1965, rock-influenced theater songs have also shunned perfect cadences. Most often this is done by using a suspension on the dominant—either a dominant chord with a suspended fourth, or a suspended chord such as the subdominant above a dominant root—rather than the dominant triad or seventh.

Suspensions used in place of triads can be seen as early as 1969 in the score of *Promises, Promises*, with music by pop composer Burt Bacharach. They can also be found throughout Stephen Sondheim's score for *Company* from 1970. Since then, final cadences with suspended dominants have become as common as perfect cadences used to be. Examples can be found in "Nothing" from *A Chorus Line*, "Not While I'm Around" from *Sweeney Todd*, "Bigger Isn't Better" from *Barnum*, "Memory" from *Cats*, "All I Ask of You" from *Phantom of the Opera*, and "The Last Night of the World" from *Miss Saigon*.

Another common rock technique is to substitute another chord for the dominant in a cadence. We saw this earlier in jazz-influenced songs, but there the substitution chord was almost always a flatted II of some kind. In rock-influenced songs, the substitution is almost always something else. One common substitution is a IV chord, which produces a plagal cadence, as in "I Got Life" from *Hair* and in the title song from *Jesus Christ Superstar*. (The plagal cadence is centuries old, but because pop songwriters of the nineteenth and early twentieth centuries avoided it, it sounds newer and fresher than a perfect cadence.) Equally common is the use of a minor rather than a major IV chord, as in "On My Own" from *Les Misérables*, and there are many other options. For example, "We Beseech Thee" from *Godspell* uses a flatted VII going to I, while "Unexpected Song" from *Song and Dance* uses a flatted III to I. Most songs that make consistent use of perfect cadences now tend to sound old-fashioned.

Harmony can either support or negate what the melody and lyrics say. Simple diatonic chords, such as major and minor triads and dominant sevenths, are usually perceived to be emotionally direct. More complex harmonies, such as diminished or augmented chords, altered chords, and extended chords such as ninths and elevenths, increase the harmonic and emotional tension to be resolved.

Another way to increase tension is to modulate to another key. Past pop hits like Stevie Wonder's "You Are the Sunshine of My Life" or Barry Manilow's recording of "Can't Smile Without You" move up a half step at the end, giving the songs a bit of an artificial "lift." This type of upward modulation can be found in musical theater as well, especially in songs by Jerry Herman and Andrew Lloyd Webber. But too many upward modulations at the ends of songs can dilute their effectiveness. When more than one or two songs in a show end this way it can become tiresome or sound arbitrary, unconnected to the dramatic needs of the story. Good theater composers know that there must be a better reason for modulating than to generate artificial excitement.

Modulation is best used for a specific purpose. Sometimes a song must modulate to accommodate two singers with different ranges, as in the duets "I've Never Been in Love Before" and "I'll Know" from *Guys and Dolls*. Both are sung by Sky Masterson in one key and Sarah Brown in another key.

A change in mode (minor to major, or major to minor) is not truly a modulation, but it can be effective in expressing shifts in emotion or perspective. One example is in "At the Ballet" from *A Chorus Line*. In the verses, which are in minor, Sheila, Bebe, and Maggie tell their stories of growing up in dysfunctional families. In the choruses, however, the three come together to describe the ballet as a place where they can escape the harsh realities of life, where "ev'rything is beautiful." The choruses are basically in major, which lightens the mood and makes an emotional contrast to the verses. Another instance is "Far From the Home I Love" from *Fiddler on the Roof*, in which the A sections are sometimes in minor and sometimes in major. We'll examine this song in more detail below.

There are two basic approaches to accompaniment. One is a chordal approach, favored by many rock-influenced composers. The other approach, which can be found in the works of Stephen Sondheim and other formally educated writers, is linear or contrapuntal, weaving together various melodic lines. The approach can vary from show to show and from song to song. It can also vary within a song itself.

Chordal Accompaniment

Contrapuntal Accompaniment

There are other options, of course. Vocalization over a purely rhythmic accompaniment can be found in the indigenous music of many cultures around the world, such as various African tribes. Musical theater composers have begun to take notice of this possibility. For example, in the "Som'thin' from Nuthin'/Circle Stomp" sequence from *Bring in 'Da Noise, Bring in 'Da Funk*, a single female voice sings to the accompaniment of tapping feet. Both the singer and dancers are later joined by an actor declaiming a monologue, adding another sonic element. Although a purely rhythmic accompaniment can be interesting, good composers use it sparingly, and only when it is appropriate to a show's setting.

Composers can change an accompaniment as a song progresses to show a change in a character's viewpoint, or simply for the sake of variety. A composer writing a song in AABA form may not want the accompaniment for the A sections to be the same each time. A song that begins with a chordal accompaniment in the first A section can include a countermelody in the second A. (A counter-melody is a second theme played at the same time as the main theme. Countermelodies are usually built using scale degrees and rhythms different from the main melody to make clear that they are distinct.) The final A section may have a completely different accompaniment. An audience may not notice these variations consciously, but they can add subliminal interest and depth.

Accompaniments also help us to recognize when a song is over. Cadences do this, of course, but a cadence may not indicate clearly enough that the song has ended. So almost all musical theater songs use a "button" to end a song. A button is a sort of musical punctuation mark; it is a single note or chord that emphasizes the song's final cadence and makes clear that the number is over.

Buttons are, as a rule, short and simple. They are usually either small and quiet or big and loud. The small button is, normally, a single low note played by a piano or *pizzicato* (plucked) by a double bass. The big button, by contrast, is either a bare octave or a full chord played by the entire orchestra.

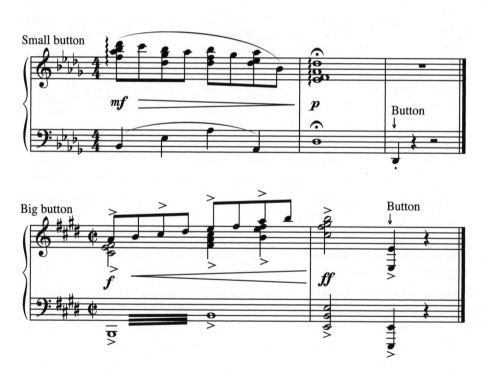

Sometimes a note or chord can be repeated for a more assertive and definite button, as in the example below.

Sometimes a song has no button, when it is intended to lead into the next number without a break. For example, in *Guys and Dolls* "My Time of Day" runs directly into "I've Never Been in Love Before."

A button makes clear to the audience that the song is over, and that it is time to applaud. Without a button the audience can become confused, unsure whether or not to applaud. The button is a long-established convention of musical theater, and audiences wishing to applaud become frustrated when a button does not appear at the end of a song.

Adding Subtext

We have already mentioned the concept of subtext in dialogue, where a character can say one thing and mean something entirely different. Composers can create subtext by using the music to contradict the lyrics. For instance, to use the major mode, which ordinarily connotes happy or positive feelings, in a serious situation sets up an emotional contradiction, and the reverse would also be true. A man singing "I'm the happiest guy in the world" accompanied by slow, low minor chords will give the definite impression that he is in denial. In "The Road You Didn't Take" in *Follies*, Benjamin Stone asserts that he has put any regrets about his life behind him—but this is contradicted by a sharp, syncopated dissonance in the accompaniment.

Variety

The amount of variety that is appropriate for a song depends on factors like the musical style and the lyrics. There are songs based largely on repetition, like folk songs and twelve-bar blues, in which a large degree of musical variety isn't necessary. In those instances it is the lyrics, the accompaniments, or a combination of the two that provides variety.

Until the 1970s theater songs often began with an introductory verse section. Because the chorus was considered the real "meat" of the song aesthetically and commercially, the verse was usually given music that was not only different, but much less interesting. (There are exceptions, of course, such as Jerome Kern's music for the verse of "All the Things You Are" from *Very Warm for May*, and Leonard Bernstein's music for the verse of "Lonely Town" from *On the Town*.) Since most theater songs now dispense with a verse altogether, this musical disparity has largely disappeared.

In most theater songs a contrasting section—the B section, release, or bridge discussed in the previous chapter—provides sufficient variety. We have already seen that lyrically a B section provides a different perspective on the subject at hand. In a sense, this happens musically as well, with a harmonic shift that either emphasizes a chord other than the tonic or temporarily modulates to another key. For example, in Cole Porter's title song from *Anything Goes*, the chorus has an AABA structure. Both the first and second A sections, which begin "In olden days" and "Good authors too," are firmly in the home key (C major in the original show). The B section temporarily tonicizes E major before moving on to E minor and then back to the original key in the final A section.

Sometimes a simple alternation between sections is enough to achieve the balance between repetition and variety. Songs that use a simple AB or verse-refrain alternation include "Oh, What a Beautiful Mornin' " from *Oklahoma!* and "Sunrise, Sunset" from *Fiddler on the Roof*.

An unusual musical structure can be found in "Far from the Home I Love" from *Fiddler on the Roof*. The form of the lyric is AABAA', which is somewhat unusual in itself, and it is enhanced by musical changes. The first and third A sections are in minor. The second A section has essentially the same melody, but it is in major rather than minor. This gives us both repetition (in the lyrics, the melodic contour, and rhythm) and variety (in the melodic and harmonic modal change). The final A section begins in major, paralleling the second section, but at the words "Who could imagine" it switches back to minor and ends in minor. This creates a "bookend" effect, concluding the song in the same mode with which it began. The alternation of major and minor also enhances the bittersweet quality of the song and the moment in the show.

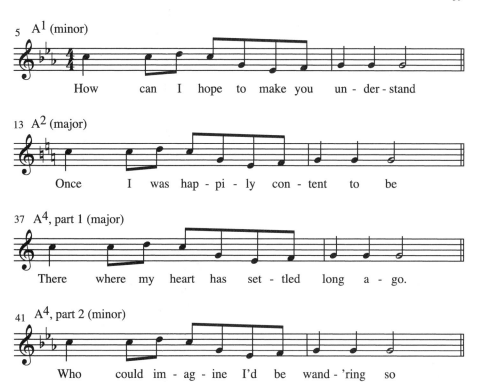

In addition to the need for variety within each song, there is also a need for contrast from one song to the next. This contrast helps prevent the boredom that can set in if there are too many similar songs in a row. While it is possible to have several ballads or up-tempo songs in a row, some element or elements usually provide contrast from one song to another. In *A Little Night Music*, Stephen Sondheim's basic concept was that the entire score would be in triple meter of some kind. Nevertheless, the songs have a wide variety of meters and rhythmic feels, such as polonaises, jigs, waltzes, and so on.

On the other hand, too much contrast from one song to the next can make a show come off as a stylistic jumble. In addition, songs too far removed from the setting or overall musical style of a show can seem incongruous and inappropriate, in the way a "heavy metal" song would seem in a musical set in the 1920s. Composers usually try to create a believable and consistent musical world within the context of the show.

Composers sometimes mix musical styles in revues, where there is often less of a need for stylistic uniformity, or in book shows in which the conflict of several generations is represented. A good example of the latter is from *Bye Bye Birdie*.

The show takes place in the late 1950s during the early days of rock 'n' roll, but Kim's parents sing "Kids," a song that would not be out of place in the 1920s or 1930s. The music makes it clear that they are out of touch with the "kids today," as they sing in a style from their own youth.

Sometimes, however, a composer will create a disparity that cannot be ignored. The musical style of the title song from *Phantom of the Opera* is drastically different from that of the other songs in the show. The operatic scenes are written in various operatic styles, while the backstage scenes use Andrew Lloyd Webber's usual soft-rock style, flavored with elements of Romantic art music. The title song, however, is a radical departure. It is in a techno-pop style, complete with a heavy backbeat on electronic drums. Many listeners find this jarring and incongruous with the style of the rest of the show.

Contrast in the score as a whole can be used to underline elements of the story. *West Side Story*'s score uses two contrasting musical styles to symbolize the conflict between the Sharks and the Jets: the cool jazz inflections of Tony and the Jets on one hand, and the hotter Latino rhythms of Maria, Anita, and the Sharks on the other. When these stylistic divides are crossed, as when Tony sings "Maria" with its flavor of the Latin habanera rhythm, a subtle dramatic point is made. The score of *Follies* is intentionally in two disparate styles, the style of Ziegfeld *Follies* songs and a more modern style. The resulting score works on two levels of reality, as the characters move between the present and the past.

Writing a show with little variety is risky. A show in which all of the songs are of the same genre or period style can lead to a feeling of predictability for the audience. A certain degree of musical consistency can be desirable, as it is in *The Boy Friend* or *Little Shop of Horrors*, but even in these shows there is considerable variety. The same is true of *A Little Night Music*, as mentioned above.

Lack of musical variety is an inherent danger with through-composed shows. In traditional book shows, the beginning of a musical number marks an important point, a heightening of excitement. Continuous music levels the differences between the big moments and the small. Unless the writers are careful, all of the music—the big numbers and the recitative transitions alike—can seem to have the same emotional weight, and thus the differences between important points and less important ones can become obscured.

VOICE LEADING, STEP MOTION, AND THE LONG LINE

Just as melodies have structure, such as an arc shape leading toward a climax, so composers try to ensure that harmonies don't just follow each other at random, but flow in coherent progressions. Even in progressions, though, skilled composers do not simply follow one chord with another; each chord progresses to the next with motion that is as smooth as possible. When each note in one chord leads smoothly to a note in the next chord, then each series of notes can be played by a different instrument as a melodic line. The technique of treating music as simultaneous lines rather than chords is called *voice leading*. This term applies whether the music is being sung by actual voices or played by instruments.

Stepwise motion is considered the smoothest motion from one note to another. A line that moves by step (which here means either whole step or half step) is also the easiest type of line for voices to sing or instruments to play. (The next easiest motion is leaping by thirds through the tones of a triad.) But beneath the surface of most good melodies and harmonies, running throughout a song, lie similar motions by step, giving the music the feeling of long lines that smoothly and inevitably rise to a climax or subside. Without going to the same extent as the German theorist Heinrich Schenker, who made a career of charting the hidden long lines in classical pieces, let's look at examples of the use of step motion and long lines in two famous theater songs, "Tonight" from *West Side Story* and "Bewitched" from *Pal Joey*. In these examples, the notes that form each "background" line are shown and connected. These constitute the "spine" of the music, without the surface tissue (consisting of other notes such as nonharmonic tones and repetitions) that usually conceals it. In the second example, notice that not only does the melody have an underlying line, but the harmonies also move by lines. Often there are several background lines running through the music at the same time.

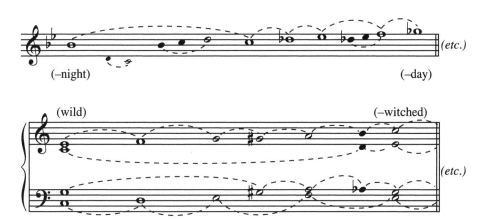

These hidden connections are not necessary to a song, and many songs have little or no evidence of a long line. But background lines often hold the music together, giving it a structural integrity and an overall shape, much as a strong spine holds a healthy body together. Stephen Sondheim has said that he always tries to think in terms of the "long line" when he composes a song.

FORMATS

The first thing a composer writes down is usually a *lead* (pronounced *LEED*) *sheet*. This is the bare skeleton of a song. Below the melody are the lyrics, with their syllables separated by hyphens, as in all vocal music. Above the staff are the chord names. Sometimes the tempo indication also gives a clue as to the musical style of the song, and important background instrumental or vocal cues may be included as well.

This Is a Simple Song

Words and Music by
Allen Cohen and Steven L. Rosenhaus

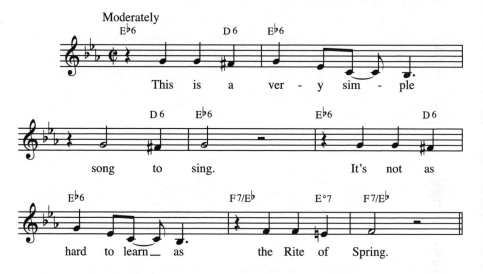

For some composers, the lead sheet is the last step in the creative process. These composers turn the lead sheets over to arrangers to flesh things out as needed. For other composers, the lead sheet is the song in its infant stage, and these writers will then create a piano-vocal arrangement or "piano chart" (see below). As mentioned in chapter 1, composers rarely write their own orchestrations. Some composers like Stephen Sondheim skip the lead-sheet stage entirely and only write piano-vocal arrangements. A piano chart consists of the singer's part—the melody and lyrics exactly as on the lead sheet—and an accompaniment arranged for piano. Here is the previous example in piano-vocal format.

This Is a Simple Song

Words and Music by
Allen Cohen and Steven L. Rosenhaus

Lead sheets give the singer and the accompanist a limited amount of information. A piano chart more closely approximates what will be heard when the music is orchestrated.

Writing for the Voice

RANGES

A theater composer has to keep an open mind when it comes to casting. No one with the type of voice for which the role was written may show up to audition, or the producers and director may have someone in mind who they feel is perfect dramatically for the part, even if that actor isn't the right vocal type. In any case, the composer may have to accept the change.

Does that mean that Tevye could be sung by someone with a higher voice? He has been, many times. Could Henry Higgins or the King of Siam be performed by a legitimate singer? They both have been. Still, composers find it a good idea to create roles musically according to vocal types, to help delineate the characters for themselves and for the audience—and to help create contrast from one character to another.

There are four basic vocal ranges. Moving from high to low, there are two female voice parts, soprano and alto, and two male voice parts, tenor and bass. The ranges shown below are those most often encountered in choral group settings.

Of course these divisions are somewhat arbitrary and limited, not taking into account voices with extended ranges—such as coloratura sopranos and countertenors—or voices with ranges that lie between those of the others. The baritone sings higher than a bass but not as high as a tenor; similarly, the mezzo-soprano sings higher than the alto but not as high as the soprano.

Note that each vocal range is approximately an octave and a half. This is the average range for a musical theater singer. Songs for children usually have narrower ranges. Sometimes a composer will write with a specific voice or singer in mind; in that case a song may have a narrower or wider range. Voices are often described in terms of their qualities as well as their ranges. Lyric sopranos and dramatic sopranos have similar ranges, but their voice qualities are different.

"Range" refers to every note, from lowest to highest, that the singer can produce comfortably, clearly, and with control. A song might not use a singer's entire range, however, and it might not use the entire range equally. The part of the range used most in a song is called the *tessitura*. The tessitura of most songs lies in the middle of the singer's vocal range, with the extremes of range saved for the climactic point or for some special effect.

Composers will often use a high or low tessitura to help a song create a particular feeling. Vocal cords vibrate more freely at the low end of a singer's range, and increase in tension at the high end. As a result, songs with high tessituras often seem more intense than songs with low tessituras, which can give the impression of being relaxed or tired.

A good example of a song with a low tessitura is "Ol' Man River," sung by the dock laborer Joe in Act One of *Show Boat*. The overall range for the song is an octave and a sixth, but much of the song lies within an octave of the lowest note. It is only in the last section, at the climax of the song, that the full range is utilized. The prevalence of low notes helps to convey Joe's weariness with life.

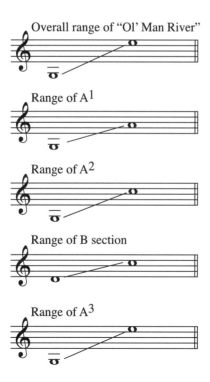

LYRICAL CONSIDERATIONS

Most consonants, like *b*, *t*, and *k*, cannot be sung on a pitch or sustained; they are produced by partially or completely blocking the path of air through the mouth with the tongue or lips. Some "soft" consonants, like *l*, *m*, and *n*, do not completely block the path of air. They can be sung sustained; humming, for example, is essentially a sustained *m*. Vowels, on the other hand, are produced with little or no obstruction of air, and are more "open" than consonants. Vowels can be sung and sustained easily; in essence they are what makes words "sing." A long sustained note is usually the most effective way to end a theater song, which explains why so many end with words like "today," "now," or "love."

Some vowel sounds, such as "ah," "oh," and "aw," are relatively open. Other vowels, including "ay," "ee," "eh," and "ih," are closed or tighter. The higher anyone sings, the more noticeable are the differences between open and closed vowel sounds. Open vowel sounds stay open at the top of a singer's range, but closed vowels are usually harder to sing and not as pleasant to hear. If the composer has set a word like "see" on a very high note, the singer may either modify the vowel to something slightly more open, like "sih" or "seh," or be unable to sing it well. (Many Italian words end with open vowels like *o* or *a*, which may be one reason why song has always been so popular in Italy.) A classic example of the bad setting of a lyric can be found in the national anthem of the United States, where a closed "ee" sound is sung on the highest note.

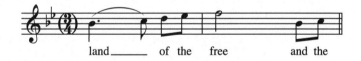

In chapter 5 we mentioned some of the factors that impede singability or comprehensibility in lyrics: one syllable ending and the next one beginning with the same consonant, too many sibilants, or awkward combinations of consonants. Sometimes a lyricist can't avoid such things, in which case it is up to the composer to ameliorate the problem if possible. With adjacent identical consonants, for instance, a composer might put a rest between the two syllables.

BREATHING

Songwriters must always have in mind a singer's need to breathe. A good composer tries to set musical phrases to coincide with lyrical ones, allowing time to breathe either between lines or in other logical places.

Like all good rules of thumb, of course, this one has exceptions. Sometimes a heightened emotion such as excitement, anger, or panic is best conveyed by setting the words so that they run on quickly without stopping. But even in such situations, the composer must still build in places for the singer to take a "catch breath."

WRITING FOR MORE THAN ONE VOICE AT A TIME

Writing music for more than one voice at a time creates a musical "texture" of vertical and horizontal relationships among voices, much as threads are crossed on a loom to create cloth. The four basic types of texture are monophony, homophony, heterophony, and polyphony. Usually a texture is considered to include all the music—accompaniment as well as singing voices—but here we will consider only the texture of the voices.

In *monophony*, only a single melody is heard. An example would be when people sing in unison (at the same pitch) or in octaves. Gregorian chant is monophonic.

Singing in octaves

Singing in unison or octaves is good for ensembles of children; it promotes confidence in their singing, helps to keep things in tune, and covers the occasional mistake. This is of particular importance when child actors play leading roles in a show, as in *Oliver!* and *Annie*. "Food, Glorious Food" from *Oliver!* is sung by a chorus of boys in unison, except for an occasional solo. Similarly, the little girls in *Annie* sing "It's a Hard-Knock Life" in unison. Adults who sing in unison or octaves have the same advantages.

Monophonic texture can also convey a commonality of deeper emotion. Moments of mutual frustration, as in "It's Hot Up Here" from *Sunday in the Park with George*, or determination, as in "Do You Hear the People Sing" from *Les Misérables*, are effectively expressed by the stark power of voices singing in octaves.

Homophony is typically used to describe a texture with a single melody that is supported by harmony in some voices. Perhaps the easiest way to think of homophony is as "melody and chords." Most church hymns, for instance, are sung homophonically, with few or no differences in rhythm from voice to voice.

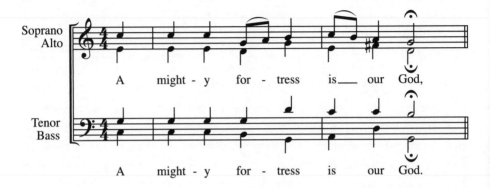

Heterophony is a texture in which the same melody is sung by two or more voices simultaneously, but with slight rhythmic or ornamental differences between them. It can be found in the indigenous music of many non-Western peoples, including those of Japan, China, and Africa. Heterophony is rare in musical theater.

Polyphony is a texture with more than one melodic line, each independent from the others and roughly equal in importance. The technique of combining these lines in a pleasing way is called *counterpoint*.

Good composers know that the trick of contrapuntal vocal writing is to make sure that the lyrics can be heard and understood. One way to accomplish this is to present the separate melodic lines individually before combining them. A venerable example of this technique can be found in a number from Irving Berlin's first complete Broadway score, *Watch Your Step*. First one character extols the virtues of old-fashioned ballads in "Play a Simple Melody." Then another character responds by calling for a new ragtime number in "Musical Demon." Both songs are then sung simultaneously, but since we've already heard them individually, we don't need to follow them carefully. We can simply enjoy how well the lines go together.

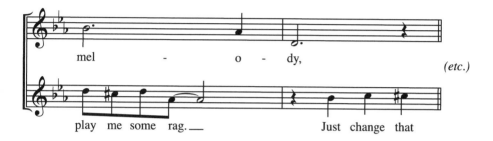

Combining two tunes contrapuntally was a favorite device of Berlin's; he used it throughout his career, in such songs as "You're Just in Love" from *Call Me Madam* and "An Old-Fashioned Wedding" from the revival of *Annie Get Your Gun*. This technique was also used by Meredith Willson in several numbers from *The Music Man*, and by Stephen Sondheim in the combination of "You're Gonna Love Tomorrow" and "Love Will See Us Through" from the Loveland sequence of *Follies*.

Another way to make contrapuntal lines intelligible is to make them not only melodically and rhythmically distinct, but isolated from each other, so that one character sings his most significant words in the "holes" of the melody sung by the other. Stephen Sondheim is a master of this, as can be seen in such songs as "Poor Baby" and the title song from *Company*.

KEYS

Normally, keys are considered flexible, and they are often transposed (changed) to accommodate a particular singer's range. This is in stark contrast to through-composed operas, which lock the singers into particular keys, because changing the key of even a short section would theoretically require the transposition of an entire act. In addition, musical theater composers often have to perform their own songs. As a result, they tend to write to suit their own voices, knowing that the keys will change. Transposing a song into another key is a relatively mechanical operation that can wait until the singer has been hired and has learned the song. Then the song can be put into the best key for that singer.

Beyond the Songs

One of the most venerable conventions of musical theater is the overture. Like the nineteenth-century opera overture from which it descended, its form is usually that of a *potpourri*, a medley that introduces some of the show's important tunes one after the other. The other primary purpose of the overture is to get audience members into their seats and quiet them down so the show can start. But the same goal can be accomplished by the convention of lowering the house lights before the show begins, and this convention has been around since Richard Wagner started it at his Bayreuth theater in 1876. An increasingly common alternative to the potpourri overture is a "prelude," which tends to be shorter and more tightly structured. A prelude usually has only a single theme, tempo, and mood; it sets the milieu or tone for the scene about to follow, for the show as a whole, or both. Examples include the "Introduction" to *Brigadoon*, the brief "Prologue" and title song of *Little Shop of Horrors*, and the preludes to each of the three acts of *The Apple Tree*. Most of Stephen Sondheim's shows since 1970 have used preludes instead of overtures; the prelude to *Company* (although it is called an "overture") and the prelude to *Sweeney Todd* are especially memorable examples.

Although overtures have traditionally been separate, stand-alone numbers, sometimes the overture or prelude leads directly into the show without a break, as in the *Carousel* "Waltz" or the overture to *My Fair Lady*. It has also become common for shows to begin without either overture or prelude; the curtain simply rises on the opening number or scene, as in *West Side Story, Fiddler on the Roof,* and *Cabaret*. Nevertheless, a big, brassy, and well-arranged overture played by a full-sized Broadway orchestra—which is an increasingly rare commodity— has an excitement and exhilaration like nothing else, and remains one of the most characteristic aspects of musical theater. Perhaps the classic model is the overture to *Gypsy*, with its evocation of the garish world of burlesque. Another wonderful example, clearly modeled after *Gypsy*'s, is the overture to *Merrily We Roll Along*.

The composer almost never writes the overture, the music for the dances, the "Bows" music for the curtain calls at the end of the show, the "Exit Music" played while the audience leaves the theater, or the brief underscoring cues of scene-change music found in many shows. These pieces are usually the very last things written, and they are usually done by an arranger or the orchestrator. Dance music requires its own special skill; the arranger has to work with the choreographer to make the music interesting and distinctive, yet integrated thematically with the rest of the score. Underscoring is most often found at the end of a scene, where it is sometimes called a "tag." Writers traditionally tried to end every scene with a song or something musical. When a song ended a scene, a tag usually followed it as soon as the applause began, to make a smooth transition to the next scene, and often to cover a scene change in between. The tag may use music from the number just ended, or for a change in mood it might use music from elsewhere in the show. Nowadays, because of the swifter pace of scene changes, these music cues tend to be shorter than in the past, and less common.

Occasionally underscoring is heard under spoken dialogue that neither comes out of, nor leads into, a song. Usually this type of nondiegetic or "incidental" music is used to intensify the emotions being portrayed, as in movies, where it is far more common than in musical theater. Act Two of *South Pacific* includes several pieces of incidental music, such as in Scenes Five and Six, when Lieutenant Cable and de Becque begin their mission.

Another effective example of underscoring is heard near the end of Act Two of *The King and I*. When Mrs. Anna receives a letter from the dying king, she changes her mind about leaving without saying goodbye to him. As she reads the letter, the orchestra poignantly reprises an earlier song about the king, "Something Wonderful." This song is also quoted at the end of the show, as the king dies and his heir Chulalongkorn becomes the new king.

The Musical World of the Show

SUGGESTING TIME AND PLACE

The majority of successful shows use the musical language of their own time, regardless of the time and place in which they are set. The reasons for this should be clear: Good composers express themselves in their own musical styles, which are necessarily of their own time. As discussed before, music written in a style that is not reasonably contemporary tends to sound hackneyed.

While deciding how to suggest the time and place of the story, the creative team must also decide the extent to which time and place need to be indicated. Composers have the options of writing within a period style, ignoring it altogether, or finding some middle ground—a modern style "flavored" with touches of a period style. Every show requires its own solution.

One factor in determining the extent to which a composer needs to suggest the time and place of the story is whether the time and place has ever existed.

A musical based on a true event such as the sinking of the Titanic, or on a fictional story with a specific location and epoch, may call for some flavor of the music of that era. On the other hand, a story set in "limbo"—with no specific time or place—gives the composer considerable stylistic freedom. Folk tales are usually set in a limbo of time more than of place; they occur in the area of their origin, but "long ago." For instance, several numbers in *The Lion King* contain or imitate African tribal music. When composers don't know the music of an era, they often research it. Even if they end up not using it directly, it will inform everything they write.

Sometimes any attempt to evoke the time and place of the story through music is futile, because the original musical styles are either unknown or would be unfamiliar and boring to a modern audience. In such cases some other style may serve the story in ways that the actual music of the time would not. Consider *A Funny Thing Happened on the Way to the Forum* from 1962. Audiences would know little or nothing about the music of ancient Rome. But they would be familiar with musical styles associated with burlesque, which flavor songs like "Everybody Ought to Have a Maid." On the other hand, many of the first theatergoers to see *The Boy Friend*, a show from 1954 set in the 1920s, had lived through the period, or had parents or grandparents who did. Music with no period flavor would probably have seemed incongruous.

How far does one go to provide a sense of time and place? It would seem logical to quote actual music from the era, as it would immediately place the listener in the proper setting. Quoting, however, detracts from the composer's role as a creator of the show. In addition, it is legally risky to quote other composers' music, unless it is in the public domain (that is, no longer under copyright). It is aesthetically risky as well. For one thing, it invites comparisons between the quoted material and the new material, and audiences usually favor what is more familiar. For another, quoting too much can have a distancing effect; the use of music from outside the world of the show can break the illusion of reality that most book musicals attempt to create.

Another approach is to write *pastiche*, music that deliberately imitates or evokes the style of a particular composer or era. An example already mentioned is the pastiche in *Follies*. Half of the score is written in the styles of older composers like Jerome Kern and George Gershwin from the Ziegfeld *Follies* era. The other half sounds like pure Sondheim.

But as with actual quotation of older music, too much pastiche tends to sound overfamiliar and dated. Occasionally a score will include a single pastiche number, and if the pastiche style is not too far removed from the style of the rest of the score, this can be effective, as in the "Side by Side by Side" number from *Company*. When the rest of the score is very different in style, pastiche songs can sometimes work as "novelty" numbers. For instance, "Herod's Song" is the only number in *Jesus Christ Superstar* not in a rock idiom. Its old English music-hall style intensifies its comic incongruity, although for some people the song seems *too* incongruous with the rest of the score.

Perhaps the most prudent approach for composers is to flavor what they write with traces of the appropriate musical styles. This is not pastiche, which is a deliberate attempt to imitate an older style, but rather composition in the composer's own style, whatever that might be, with little touches that evoke the time and place in which the show occurs. Richard Rodgers was a master of this, as in *The King and I*, *South Pacific*, and *Flower Drum Song*, all of which have flavors of the "mysterious East."

These touches of style can be almost anything that conjures up an association with the show's time and place: a particular melodic interval or mode, a distinctive rhythm, or a type of harmony associated with the setting. In *The King and I*, Richard Rodgers uses his normal musical language for all of Mrs. Anna's numbers. For anything to do with the King of Siam or his family or subjects, however, the accompaniments make extensive use of open fourths and fifths, often moving in parallel, as in "A Puzzlement." Composers have often used fourths and fifths to evoke a clichéd "Oriental" sound.

Sometimes rhythmic style can help set the milieu of a show. Specific dance rhythms often prove useful in this regard. A Charleston rhythm would suggest a setting in the 1920s; a bolero rhythm would evoke Spain; a Calypso rhythm would imply a West Indies setting. Such indications are also dependent on their context, but it's clear that rhythm can be valuable in setting a scene. Specific instruments or timbres are also often used to help suggest a time and place. When composers use ethnic instruments sparingly, it is often enough to put the audience in the proper frame of mind, as the Kabuki instruments help to set the milieu at the beginning of *Pacific Overtures*.

THEATER MUSIC AND POPULAR STYLES

Since 1955, rock-style pop music has steadily grown in popularity, becoming more and more the *lingua franca* of theater songwriters. Even rap music, which often has no singing at all, has found its way into musicals as traditional as Sondheim's *Into the Woods*—although it could be argued that having a character speak rhythmically over music goes back at least to 1957, the year of *The Music Man*.

As discussed earlier, the primary difference between theater songs and pop songs is that theater songs serve a dramatic function. Pop songs, on the other hand, are often written without specific characters, locales, or stories in mind. There are musical differences as well.

Contemporary pop music doesn't need a good melody to be memorable. Many hit songs have succeeded because of a striking title, an interesting harmonic progression or accompaniment "riff," an unusual percussion rhythm, an arresting vocal quality, or simply the popularity of the performer. Songs like Chuck Berry's "Johnny B. Goode" from the 1950s or Bob Dylan's "Like a Rolling Stone" in the 1960s were popular, but not because their repetitious melodies were well constructed.

Theater songs have to move the story and characters along in their physical, emotional, or intellectual journeys. The melodies for theater songs tend to be constructed with the "long line" in mind; this usually parallels the character's or story's journey. Musical theater songs have a climax, while pop songs rarely do.

When Rose sings "Everything's Coming Up Roses" at the end of Act I of *Gypsy*, for example, she is not just singing a song about how fabulous things are about to be. Her world has essentially come crashing down around her, but she sings the song as both denial and defiance. At the climax of the song she sings that her lucky star is due, and the word "due" is set on the climactic note of the song. This underscores her determination to change her fortunes. Pop songs, on the other hand, have been called "three-minute stories," for they create their own self-contained worlds. Except in "concept albums" like those of Pink Floyd, pop songs do not tell a larger story or help flesh out a particular character; the journey, if there is one, occurs only within the song itself. As a result, pop songs usually do not have musical climaxes, even when they tell a story, as in the Beatles' "She's Leaving Home" or Gordon Lightfoot's "The Wreck of the Edmund Fitzgerald."

There are other differences between musical theater songs and pop music. Pop songs are almost always written in one of the styles currently in vogue, while theater songs tend to be more eclectic, incorporating diverse styles and techniques. With some prominent exceptions, pop music tends to be less adventurous harmonically than musical theater songs. Pop songs are usually about three minutes long, but theater songs can be shorter or longer, depending on the needs of a particular scene. Another difference is that many pop songs have improvisational solos; in some types of rock, these solos are the longest part of the song. Any instrumental interludes in a musical must be timed to the action.

As broad as the spectrum of pop styles seems, it is narrow compared to what is available to composers of musical theater. Any style that helps a composer tell a story can be used, whether it comes from pop, jazz, folk, classical, or ethnic music. Canny theater songwriters make themselves familiar with a wide variety of music, and everything they hear informs what they compose. But they also remember that the requirements of theater music are specific; the music must reflect the characters, the story, and the setting, as well as the writer's own background.

As the above discussion should make clear, theater and pop songwriting are fundamentally different, not so much in musical idiom as in their natures and functions. Some very successful pop songwriters have written musicals that failed because of their refusal to acknowledge these differences. Duke Ellington wrote *Beggar's Holiday* as a jazz version of *The Beggar's Opera*, but it failed. More recently, critics agreed that the score for Paul Simon's show *The Capeman* was wonderful, but only as pop music; as theater music it was a failure. Nothing prevents talented songwriters of any kind from writing good musical theater, but this can only happen when they understand the nature and requirements of the genre.

Musical theater can incorporate almost any popular style successfully. Popular rhythms, sounds, and styles work best in theater when they are integrated into well-written theater songs rather than inserted whole. Two examples of good musical theater scores that successfully incorporate pop styles are *Little Shop of Horrors*, which uses rhythm and blues, rock, and doo-wop styles from the 1950s and early 1960s, and *Once on This Island*, which uses Caribbean styles.

UNIFYING THE SCORE

Most good shows have a general consistency of tone, in part because good composers write music in their own styles. Artistically ambitious composers use other means to further unify a show. Scores can be integrated both motivically and thematically. Sometimes a motif permeates the songs of a particular character or group of characters, unifying things in a subliminal way. In *West Side Story*, for example, Leonard Bernstein uses a group or "cell" of three notes—the tonic, the raised fourth, and the fifth of a scale—in a variety of ways. This cell can be found in almost every number of the score. Here are examples of its use from the Prologue, "Something's Coming," "Maria," and "Cool."

Stephen Sondheim uses the motif of a falling third followed by a falling second to unify *Company*. The basic motif is sung to the words "Bobby, Bobby" by offstage voices at the beginning of the Overture. The motif is also expanded into a theme, which we hear as the verse of the title song. After that the idea returns often, either as the motif or as the verse. "Side by Side by Side" opens with an instrumental introduction based on the motif, and it appears in the middle of the song as well; the motif also appears in the accompaniment that begins "Poor Baby." The verse reappears before "Have I Got a Girl for You?" and "Being Alive," and the middle section of "Poor Baby" quotes both the verse (at "Robert . . . Bobby . . .") and the motif (at "You know, no one").

Some composers use a "motto theme" to help unify the score. A motto theme is a distinctive recurring phrase. In *Gypsy*, Momma Rose has a motto theme, "I had a dream," which is first heard in the song "Some People." This returns at significant points in the show, most crucially in her breakdown during "Rose's Turn." *Gypsy*'s lyricist, Stephen Sondheim, later used motto themes in various ways in his own scores. In *Follies*, the phrase "Hey up there, way up there" is used as a motto.

An unusual use of themes can be found in *Wonderful Town*, in which Leonard Bernstein repeatedly uses a section of one song for a different section of a later number, creating chains of thematically linked songs that tie the show together musically and dramatically. In the verse of "What a Waste," magazine editor Robert Baker brusquely tries to convince Ruth Sherwood that she will never make it in New York City. The beginning of the verse quotes "Ohio," a song Ruth has just sung with her sister. In the chorus, the A section tells Baker's story ("Born in Duluth"). Later, the same melody appears as the verse of "Pass the Football," this time to tell the story of the Wreck, who sailed through college because he was a star quarterback. At the end of Act Two, Baker realizes that he has fallen in love with Ruth. The verse of "What a Waste" has now become the A section of his song "It's Love."

Meter can also be a unifying factor, as in Sondheim's use of triple meter in all its permutations throughout *A Little Night Music*. Sometimes even the absence of a particular time signature can subtly help unify a score. In *Do I Hear a Waltz?*, Leona Samish waits to hear an imaginary waltz that will signify that she has found true love. Composer Richard Rodgers avoids three-quarter time until midway through the second act, when Leona finally hears her waltz.

To sum up: A play is an illusion of reality, and anything incongruous has the potential to destroy that illusion. In musical theater the illusion is every bit as fragile and is even harder to achieve, as the audience has to accept the convention that characters sing to express themselves rather than talk. Thus, like the lyrics, the music must be appropriate to the story and the characters. For a song to work properly, every aspect of it must suit the characters singing it, the emotions inherent in the specific moment, and the setting and tone of the show.

PART II

WRITING A SHOW

CHAPTER 7

GETTING STARTED

In part I of this book we surveyed the fundamental principles and techniques underlying the creation of musical theater. Once you have grasped the basic techniques and principles, the best way to learn to write musical theater is to *write*, and then to present what you have written to an audience of some kind. No book can provide a substitute for that, but in this part of the book we can help you get started. We will use a variant of the tutorial approach, in which you can follow us step-by-step through the initial stages of writing a new musical. We will begin two musical theater projects; you can "look over our shoulders" to see how we proceed and, if you like, second-guess the choices we made. Because the processes of writing adaptations and originals are so different, our "tutorials" will include one project of each kind. The point of this approach is to reduce the anxiety of starting a new show from scratch, by breaking the process down into stages and showing how it can be done.

Of course, every writer works differently, every project is different, and there is no such thing as a single correct approach. As you read the following chapters, we encourage you to stop occasionally, think about our choices, and consider whether you would have made different ones. These tutorials will show you how two particular writers proceeded with two particular projects.

We want to emphasize that both projects were created for this book. They are not shows that we ever brought, or tried to bring, to fruition; they have not been chosen or written with an actual production in mind. Therefore all of the crucial later stages of creating a show are missing: rewriting and polishing it, seeing and hearing it on a stage, presenting it to an audience, and doing more rewriting and polishing. The only way to go through those steps is to actually put on your show in some fashion, which can be anything from a reading to a full commercial production.

Sample Project A: An Adaptation

THE STORY

For both our projects we wanted to write book musicals rather than revues, dance shows, or "cover" musicals utilizing old songs. (The ideas for these other types of shows rarely originate with the writers.) We looked for stories with the potential for a dramatic structure and a central theme or spine; stories with conflict, action, characters to care about, situations emotional enough for the characters to sing, and perhaps room for spectacle and dance. For our adaptation project, we wanted a property with which many readers would already be familiar, and which other readers could easily find if they wished.

At an early point we thought of using the 1939 film *Mr. Smith Goes to Washington*. This comedy-drama, written by Lewis R. Foster and Sidney Buchman and directed by Frank Capra, concerns a young and naive scout leader who is chosen to fill a vacant U.S. Senate seat by a cynical and corrupt political boss. We quickly decided that the film's negatives outweighed its positives. On the positive side, it remains well regarded by critics and enjoyable to watch; it has a fast pace and an interesting, even exciting plot; and there is an emotional aspect as well. But there are many negatives. The plot is hopelessly farfetched: Despite the enmity of a powerful political machine and most of his fellow senators, Mr. Smith triumphs over the political boss by quoting the Constitution during a filibuster. The climactic sequence, in which thousands of boy scouts from Smith's home state manage to distribute their little homemade newspaper to the public despite the attempts of brutal thugs to stop them, is shown by a series of montages that would not work onstage, and would have to be replaced by a good deal of talk about offstage events. Even if the production skimped on the crucial location of the Senate chamber, by its very nature the show would require a huge cast, with dozens of senators, reporters, and children. In addition, most of the strong emotion in the show comes from pride in our country's political and legislative processes, and in the light of recent political events we found it a little hard to summon up the comparatively innocent optimism of 1939. Last but not least, *Mr. Smith Goes to Washington* was produced by a major film studio, and presumably is still under copyright. Because obtaining the rights for material owned by a Hollywood studio is usually either prohibitively expensive or simply impossible for unknown writers, we decided to choose only from among sources in the public domain.

We briefly considered, and quickly rejected, Herman Melville's novel *Moby Dick*. We considered it because it is in the public domain, it is well known and celebrated, and it would be an unusual and provocative choice. The problem is that all the things that make it unusual are good reasons not to make it into a musical. For most of the book there is very little external action, and what action there is involves the hunting and slaughtering of whales. This would be difficult to represent convincingly onstage, and might be repugnant to modern audiences who know that whales are gentle, intelligent mammals. Also, while Captain Ahab is an extremely vivid character and there are a number of others, there are no female characters. In the course of an evening of musical theater, a large cast with no female voices might become monotonous. Finally, *Moby Dick* is a classic novel, one of the greatest ever written by an American; making it into a musical would add nothing, and remove most of what makes it great. It would unquestionably be a "why musical."

We eventually chose Mark Twain's short novel *The Prince and the Pauper* which, we felt, has many things to recommend it. The story, of a coincidental resemblance between King Henry VIII's son Edward and a young beggar boy, and how they inadvertently switch places, is one that generations of children have read, and it is in the public domain. The plot contains most of the essential elements for a good musical: action, dramatic conflict, an overall dramatic arc, and plenty of variety. There is also emotion inherent in the story, such as the sufferings of the young Edward when he is misunderstood and tormented, and the panic of the young ragamuffin Tom when he is treated as the Prince of Wales and then King of England.

Nevertheless, every project has its negative aspects, and *The Prince and the Pauper* has its share. The story is a fairy tale, with a number of coincidences that border on the preposterous. While children may accept the story, contemporary adults would probably find it much more difficult to swallow. Much of the book is sociopolitical satire of a way of life that has been obsolete for hundreds of years. Also, like *Mr. Smith Goes to Washington*, *The Prince and the Pauper* calls for a large cast, including both the English royal court and streets full of London's poor. Several actors and members of the chorus could play multiple secondary roles, however, and not every character in Twain's story needs to appear. In addition, unlike *Moby Dick*, *The Prince and the Pauper* has a variety of roles for both females and males. Women do not play a major part in the original story, but this could be adjusted to some extent.

Another problem with basing a musical on *The Prince and the Pauper* is the casting of the leads. It needs two boys who not only look ten years old, but resemble each other enough for the audience to accept them as identical. Most likely they would both have to act and sing well, and perhaps to dance. In many professional productions a third boy would also have to be cast as an understudy for both roles, and he would have to resemble both the lead actors. Nevertheless, there have already been several dramatic and musical versions of the book, so it can be done.

The fact that musical versions of this story have already been written, of which at least one briefly played on Broadway, would seem to be a negative as well. We decided it was not, for two reasons. First, since we were not writing the show for production, we did not need to worry about an audience being overfamiliar with the material. Second, of the previous versions in recent decades, none has been successful with the general public.

There is also the problem of language. Mark Twain's characters speak in what seems like authentic sixteenth-century English. Much of it would be off-putting and difficult for a modern audience to understand, especially in lyrics, which go by quickly.

In the end, none of these negatives outweighed the positives we found in the original, at least for our didactic purposes. Following the principles discussed in chapters 3 and 5, we decided to use modern English, with adjustments made to suggest the time and place of the story. And following the principles of chapter 6, we decided that the music should likewise be contemporary, with perhaps a slight period flavor. Finally, we decided that it would be a traditional book show with dialogue rather than a through-composed one, simply because we both preferred it.

There is one more consideration that is not necessarily either positive or negative. Because much of the novel is concerned with the grim underside of sixteenth-century English life, it can hardly be adapted as a musical for young children without eliminating a large portion of the story. On the other hand, with its fairy-tale premise, it is not truly an adult story either. So we have consciously conceived our show as a "family" musical, as discussed in chapter 3. We are aiming it at an audience of older children and their parents, analogous to the "young adult" category of readers to which the novel is usually marketed. While there are many theaters that would not be interested in this type of material, that is not necessarily an argument against it. There have been quite a few Broadway musicals aimed at this audience, or an even younger one, in recent years.

THE CHARACTERS

The Prince and the Pauper came with many characters already in place, some from British history and some from Mark Twain's imagination. We could do research on the historical figures; this would help us understand and write for our characters, particularly Prince Edward and, to a lesser extent, his sister Elizabeth. But the number of characters in the novel is huge, and we needed to trim the cast size.

Some of Twain's characters are expendable. For example, the grandmother of our pauper, Tom Canty, appears only briefly, and all she does is beat Tom, just as his father does. So we eliminated her as unnecessary.

Tom has a pair of older twin sisters, Bet and Nan. The twins are a "red herring"; their existence has no thematic or structural relationship to the resemblance between the two boys, and it seems absurd to have two pairs of twins in one story. In addition, the sisters appear only briefly in the novel, and do almost nothing. Still, a sister for Tom would be a fellow victim of their drunken father, and she could help Edward when he is mistaken for Tom. With few significant female roles in the novel, we thought that it made sense to compress the two sisters into one. This would eliminate the extraneous twin, but still provide a female role.

Henry VIII is a famous historical figure, a powerful man known for his huge appetites for both food and women. But in this story Henry appears only a few times before he dies; essentially he is a minor character. For this larger-than-life figure to appear in Act One would be another red herring, shifting the focus away from the boys and from the dashing Miles Hendon, where it belongs. In addition, many people in the audience might be bothered by any disparity between their knowledge of the real Henry and our portrayal of the character. We decided to keep him offstage, where everyone can envision their own Henry VIII.

Another character to be considered is the insane, homicidal hermit who, about halfway through the novel, takes Edward in and then tries to murder him. While the scenes with the hermit add suspense, they do little to move the story forward, and the latter half of the novel is already crammed with episodes and incidents. We decided to eliminate him and replace his scenes with new ones using more important characters. But Humphrey Marlowe, the prince's whipping-boy, gives Tom information that allows him to keep up his role as Prince Edward, and also furnishes some of Twain's most ironic humor. Humphrey seemed worth keeping, especially because he could also understudy both of the lead roles.

Below is a chart comparing the primary characters in the Twain novel and in our show. You may notice that Mark Twain never gave Tom's mother a first name, although she is one of the few female characters in the novel. We wanted to increase her importance in our adaptation, and we have given her a name.

Characters in the novel	*Characters in the musical*
Tom Canty	Tom Canty
John Canty	John Canty
Mrs. Canty	Mrs. Alice Canty
Grandmother Canty	(*eliminated*)
Nan Canty	Nan Canty
Bet Canty	(*eliminated*)
Father Andrew	Father Andrew (from ensemble)
Prince Edward Tudor	Prince Edward Tudor
King Henry VIII	(*eliminated*)
Elizabeth Tudor	Elizabeth Tudor
Lady Jane Grey	(*eliminated*)
Miles Hendon	Miles Hendon
Sir Hugh Hendon	Sir Hugh Hendon (from ensemble)
Mary Tudor	Mary Tudor (from ensemble)
Humphrey Marlowe	Humphrey Marlowe (from ensemble)
Edward Seymour, Earl of Hertford	Edward Seymour, Earl of Hertford
William, Lord St. John	William, Lord St. John (from ensemble)
Hugo	Hugo (from ensemble)
The Hermit	(*eliminated*)
Edith Hendon	Edith Hendon
poor people, courtiers, members of Canty's gang, royal servants, etc.	poor people, courtiers, members of Canty's gang, royal servants, etc.

Notice that we have cast Nan, Father Andrew, Hugh, "Bloody" Mary Tudor, Humphrey, Lord St. John, and Hugo from the ensemble; although essential to the story, they would all be small parts. Not counting the ensemble, there are five important male roles and four important female roles, and plenty of smaller speaking parts. We might decide to change some of these choices as we worked on the show, but in the meantime we had a cast list with which to work.

A useful next step is to create profiles of the main characters, as discussed in chapter 4. One interesting aspect of *The Prince and the Pauper* is that some of Twain's historical characters differ from their real-life counterparts. Usually we have remained more faithful to Twain's people than to the real ones, but we have taken a number of liberties and filled in many details that seemed useful for our dramatization. Below are short character profiles; for some secondary characters we give brief sketches rather than complete profiles.

Tom Canty. Tom Canty is ten years old, and looks identical to Prince Edward. The unwanted second child of John and Alice Canty, he lives with them and his sister Nan in Offal Court, a filthy tenement near London Bridge. Tom has a hard life. He begs on the streets by day, and he often goes hungry and is beaten by his father at night. Although the rest of his family is illiterate, Tom's innate intelligence has been encouraged by Father Andrew, a neighbor who teaches him reading, writing, and a bit of Latin. Father Andrew's tales of chivalry and aristocratic life furnish Tom with material for his daydreams, and he likes to play at being a prince. He also has a strong sense of right and wrong.

John Canty. John Canty, Tom's father, is in his late thirties, but like most poor people of the time, looks older. He is a thief by trade and a brutal, violent person by nature. He is also a drunkard, which exacerbates his abusiveness. He is big and strong.

Alice Canty. Alice is in her thirties but looks older, thin and somewhat haggard. She is very loving and protective of her children, but she is abused by her husband and is timid around him.

Nan Canty. Nan is fifteen and, like her mother, thin and haggard. She is also timid and quiet.

Edward Tudor, Prince of Wales (later King Edward VI). Ten years old, he is the son of King Henry VIII and Queen Jane Seymour, the latter having died giving birth to him at Hampton Court. Coming from an extremely privileged background, Edward is well educated. He knows Latin, Greek, French, and Italian. (The historical Edward also played the lute and was interested in astronomy, politics and stagecraft. None of these additional facts appears in the novel.) Edward is intelligent, inquisitive, and determined. Because of his upbringing, he is self-centered, frank, and accustomed to getting his own way. (In the course of the story he develops a profound sense of justice, although this conflicts with historical facts. The real Edward allowed his uncle Edward Seymour to be executed on a dubious charge.)

Elizabeth Tudor. Elizabeth (who eventually became Queen Elizabeth I), the prince's half sister, is about fifteen years old. She is poised and intelligent. She is also caring, thoughtful, and protective of her brother.

Miles Hendon. Miles is the son of Sir Richard Hendon of Hendon Hall in Kent. He is a youthful and vigorous man in his early thirties, tall, trim, and muscular. Practically penniless upon his return to England from foreign wars, Miles has fine clothes that have become shabby and threadbare, but he wears them as though they were new. He carries himself with a touch of bravado, and he knows how to use a sword. (Errol Flynn, one of the great cinematic swashbucklers, played Hendon in the 1937 movie adaptation.) Like most young nobles of the time, Miles is not especially learned, but he is quick-witted and has a keen sense of humor. He is also naturally protective of the innocent and the helpless.

Edward Seymour, Earl of Hertford. Seymour is Prince Edward's uncle and an advisor to King Henry; he is forty-seven, but looks older.

Edith Hendon. Edith is in her late twenties. She is very pretty and reasonably intelligent, but she is also fearful, weak, pale, and melancholy.

Secondary characters. Father Andrew is an old Catholic priest who, like most English priests after Henry VIII broke with the Roman Church, is very poor. He is warm and sympathetic, and fond of Tom. He is thin, with venerable white hair. Sir Hugh Hendon is a year or two younger than his brother Miles and a year or two older than Edith. He is perhaps smaller and less athletic than Miles. Mary Tudor (who eventually became Queen "Bloody" Mary I) is thirty-one years old, intelligent and educated but dour and severe. Humphrey Marlowe, the prince's whipping-boy, is about ten years old. (In the novel he is twelve, but we want him to serve as understudy to the leads.) William, Lord St. John, a lesser noble in attendance on Prince Edward, is middle-aged and sympathetic. Hugo, a member of John Canty's gang of ruffians, is in his late teens or early twenties, lean and loutish.

THE LIBRETTO

In chapter 3 we compared synopses of *A Little Night Music* and *South Pacific* with synopses of their original sources. Now we'll do a similar comparison for *The Prince and the Pauper*, setting a synopsis of Twain's story against the scenario of our adaptation.

THE PRINCE AND THE PAUPER
by Mark Twain

Chapter I, "The Birth of the Prince and the Pauper." Two boys of very different circumstances are born in sixteenth-century London. One is Tom, an unwanted child of the poverty-stricken Canty family. The other is Edward, the son of King Henry VIII and the much-awaited heir to the English throne.

Chapter II, "Tom's Early Life." Ten years have passed. Tom Canty lives in poverty in Offal Court, a foul little pocket near London Bridge. His father, John Canty, is a thief who forces Tom and his older twin sisters, Nan and Bet, to go out every day and beg. His father and grandmother beat Tom if he doesn't bring home enough money. Canty beats his wife as well. The oasis in Tom's life is Father Andrew, a neighbor who teaches the boy reading, writing, and a little Latin. Father Andrew also regales young Tom with stories of kings and princes. The stories serve as an escape for Tom, who dreams of princes and of being a prince himself.

Chapter III, "Tom's Meeting with the Prince." Tom, wanting to see a real prince, finds himself at the gate of the royal palace in Whitehall. The crowd pushes him against the gate, and a guard roughly thrusts him back. Prince Edward sees this, reprimands the guard, and has Tom admitted into the palace. The boys compare their lives, and each wonders how it would feel to live like the other. On a whim, Edward suggests they switch clothes; they realize that they look alike. Edward notices a bruise on Tom's arm from the guard's mistreatment, and rushes out to give the guard another reprimand; but the guard sees only a ragged beggar boy, and throws him off the palace grounds. Edward protests that he is the prince, but no one believes him. The crowd taunts him and hustles him away from the palace.

Chapter IV, "The Prince's Troubles Begin." Edward finds himself alone and lost. He comes to Christ's Church, a royal home for poor orphans. When he demands admittance as the Prince of Wales, a group of orphans mocks and beats him, leaving him bleeding, dirty, and faint. Near Offal Court, John Canty finds him and assumes that he is Tom. When Edward insists that he is the Prince of Wales, Canty thinks his son has gone mad and drags him home.

Chapter V, "Tom as a Patrician." Back in the prince's rooms, Tom is at first unaware that anything is amiss, but when Edward fails to return, he panics. Lady Jane Grey comes to meet her cousin. Tom tries to reveal his identity to her but she flees, taken aback by his bizarre behavior. Word spreads quickly that the prince has gone mad, and King Henry VIII, Edward's father, issues a decree that none speak of the matter on pain of death. Tom is taken into the presence of the king, where he again asserts his identity. Henry tests him and Tom is able to answer some of his questions. Concluding that his son's madness is only temporary, Henry orders that he be officially installed as Prince of Wales the next day. He also orders the Duke of Norfolk's execution. Tender-hearted Tom pleads for Norfolk's life to no avail. As Tom is led off, he realizes that he is effectively trapped in the palace.

Chapter VI, "Tom Receives Instructions." In the prince's chambers, the Earl of Hertford and Lord St. John advise Tom to hide his "infirmity" and to relearn what is expected of the Prince of Wales. Tom rapidly learns from his mistakes during a visit by Princess Elizabeth and Lady Jane Grey. Then he is allowed to rest, and Hertford and St. John privately discuss the situation. St. John is troubled by Tom's claim of not being the prince. Hertford reminds St. John that such talk is treasonable, and they agree to drop the subject. Later, Hertford contemplates his own doubts.

Chapter VII, "Tom's First Royal Dinner." After a rest, Tom is dressed by servants and conducted to an ornate dining room. He dines with the assistance of still more servants. He has not completely unlearned his old eating habits—he blithely drinks from the fingerbowl—but everyone treats his behavior as normal. Later, in Edward's room, he finds a book on court etiquette and starts to read it.

Chapter VIII, "The Question of the Seal." King Henry is adamant that the Duke of Norfolk be executed, but he is too weak to go to Parliament and sign the decree himself. The Lord Chancellor may do so by using the Great Seal, but no one recalls its whereabouts until the Earl of Hertford remembers that Henry had given it to Prince Edward for safekeeping. Hertford asks Tom about the seal, but Tom has no memory of it. (Hertford does not describe it to him.) Henry demands that the Lord Chancellor use a small copy of the Seal, and Norfolk's execution is set for the next day.

Chapter IX, "The River Pageant." That evening, the celebration honoring the installation of the Prince of Wales is held. It begins with fifty gaily decorated state barges sailing down the Thames. Amid the full panoply of a royal court, dressed in gorgeous clothes and regalia, Tom Canty emerges from the palace.

Chapter X, "The Prince in the Toils." As the pageant begins, John Canty is dragging the still-struggling Edward into Offal Court. Canty loses patience and begins to strike Edward with his cudgel, but someone stops him. Canty strikes the meddler instead and drags the boy home. Tom's mother and sisters see that "Tom" is not himself; they plead for mercy, but Canty beats the boy. Then he douses the light and orders everyone to go to sleep. But Mrs. Canty is worried about her son's madness and troubled by something else as well, an indefinable strangeness about him. She decides to test him by making a sudden loud noise and shining a light in his eyes; if he responds as Tom would, then he is her son. He does not, and she begins to suspect that he has been telling the truth. Later, with sounds of celebration coming from the street, a neighbor knocks at the door and tells Canty that the man he struck earlier was Father Andrew, who has died from the attack. Canty decides that the family must leave immediately to avoid the police; if they are separated, they are to meet him at London Bridge. They leave quickly, Canty dragging Edward. But as they reach the streets, filled with people celebrating the installation of the Prince of Wales, Edward manages to pull loose and run off.

Chapter XI, "At Guildhall." The flotilla sails down the Thames to the City, where the celebration for the Prince of Wales continues in the Guildhall. Edward demands entrance to no avail, and he is taunted and threatened by the crowd. Miles Hendon, a handsome nobleman down on his luck, steps in to aid the youngster. Someone tries to grab Edward, but Miles strikes the man down. The crowd attacks both Miles and the boy, when suddenly a king's messenger breaks through to announce that King Henry is dead. Inside the Guildhall, acclaimed as the new king, Tom immediately pardons the Duke of Norfolk.

Chapter XII, "The Prince and His Deliverer." Edward and Miles elude the crowd and make for Miles's lodging, an inn on London Bridge. Outside the inn they run into John Canty, who claims the boy but backs off when Miles threatens him. Up in Miles's room, Edward falls asleep. Miles decides to bring the boy to Hendon Hall, his ancestral home in Kent, where his family will welcome them. When Edward awakens they eat, while Miles explains that he served in the continental wars for three years, then was captured and spent seven years in an enemy prison. He has just returned to England. Edward tells his story; Miles thinks the boy is mad, but kindheartedly shows him the deference he demands.

Chapter XIII, "The Disappearance of the Prince." The meal finished, Miles and Edward sleep. At dawn, Miles takes Edward's measurements with a piece of string; he leaves for a short time and returns with a used set of clothes for the boy, but Edward is missing. As a servant enters with breakfast, Miles questions him and learns that while he was out, Edward was given a message, purportedly from Miles, to meet him at the southern end of the bridge; when Edward left the inn he was followed by a man who, from his description, sounds like John Canty. Miles sets off to find him.

Chapter XIV, "Le Roi est Mort—Vive le Roi." It is the morning after the celebration. Tom wakes with the hope that he has been dreaming the last few days, but he is still in the palace, and is now the King of England.

(*Since we will not be taking our adaptation further than this point, we summarize the rest of the novel very briefly*).

Chapters XV through XXXIII. As the new king, Tom holds an audience and attends to other duties. His native intelligence allows him to make fair and perceptive judgments in several legal cases brought before him. In the evening there is a state dinner in his honor; this time Tom gets through the meal flawlessly.

Meanwhile, Miles searches for Edward, who has been abducted by John Canty. At a meeting of Canty's gang of ruffians and vagabonds, the prince learns of the darker side of English justice. He proclaims his identity, but the gang mocks him. He is assigned to work a con game with a young gang member named Hugo, but his refusal to lie ruins the swindle, and he escapes in the ensuing confusion. The next day he is fed by a kindly peasant family, but he has to run away when Canty and Hugo show up. He is taken in by a hermit who accepts Edward as the king, but only because he himself is insane. He ties Edward up and is about to kill him when Miles appears outside. The hermit leads Miles on a wild-goose chase, leaving Edward trussed up, where Canty and Hugo find him.

Back with the gang, Edward's honesty proves such a liability that Hugo frames him for stealing a pig. Miles arrives just as Edward is arrested. The boy is found guilty, but Miles cleverly bluffs the guard into letting the boy escape. They travel to Hendon Hall, where Miles learns that in his absence his father and older brother have died; his rascally younger brother Hugh has assumed their father's title and married Miles's fiancée Edith. Miles has been declared dead, and everyone is too afraid of his brother to admit that they recognize him. He and Edward are arrested and thrown into prison. There they learn that "Edward the Sixth" is about to be crowned king. When Miles is sentenced to sit in the pillory, Edward protests, and Sir Hugh suggests that he be whipped. Miles offers himself in Edward's place and receives a dozen lashes before being put in the stocks. Miles decides to appeal to the new king for justice, but when he and Edward arrive in London, they are separated by the crowds that throng the streets for the coronation.

Meanwhile Tom has grown somewhat accustomed to the royal life. The day of his coronation begins with a procession through London. Along the way Tom's mother recognizes him for who he really is and pushes through the crowd to him, but he denies that he knows her. He is instantly ashamed, but she has already disappeared in the crowd.

The coronation proceeds at Westminster Abbey. The Archbishop of Canterbury is about to put the crown on Tom's head when Edward, who has managed to enter the cathedral, stops everything by declaring himself king. Tom stuns the crowd by agreeing. The attendants realize that the two boys look identical, but Hertford demands proof that Edward is who he claims to be. One question can be answered only by the true Edward: Where is the Great Seal? Edward tells them where he remembers leaving it, but it is not there. The confusion is resolved when Hertford briefly describes the Seal. Now knowing what it looks like, Tom reminds Edward that he had hidden it immediately before leaving his room. Edward then remembers where he actually left the Seal, and the rightful king is crowned.

Miles arrives at the palace after the coronation, bearing a letter that Edward had written to prove himself king. When Miles is seized as another pretender to the throne he expects the worst, but he is led in to see Edward, whom he thought mad, now sitting on the throne. Edward rewards Miles, making him the Earl of Kent; strips his brother Hugh of his title and estates; and gives Tom his royal protection and the title of King's Ward. Hugh Hendon, though not prosecuted, flees for the continent where he soon dies, and Miles finally marries his Edith. John Canty is never heard from again. King Edward VI reigns briefly, but with justice and compassion.

THE PRINCE AND THE PAUPER
A Musical in Two Acts

Act One. Scene One. London, 1547. The streets of London near the inner City. We are introduced to the poor people of Offal Court, including Tom Canty, his family, and Father Andrew. Tom sees Prince Edward and his retinue pass by on the way to the royal palace and follows them.

Scene Two. Outside the palace. Tom pushes his way through the crowd to catch another glimpse of the prince. Thrust against the gate, he is treated roughly by a guard. From a window the prince sees this, reprimands the guard, and commands that Tom be brought inside.

Scene Three. The prince's chambers. The boys compare their lives, and each wonders how it would feel to live like the other. At Edward's suggestion they switch clothes, and they realize that they look alike. Edward notices a bruise on Tom's arm, and rushes out to give the guard another reprimand.

Scene Four. Outside the palace. Edward comes out and starts to chastise the guard, but the guard sees only a ragged beggar boy, and throws him off the palace grounds. Edward protests that he is the prince, but no one believes him. The crowd taunts him and shoves him away from the palace, until he lands in the hands of John Canty, who drags him off.

Scene Five. The prince's chambers. Tom preens in the prince's clothes for a while, but soon begins to wonder what happened to his new royal friend. He looks outside the room, where he is instantly met by armed guards; he panics and jumps back into the room, slamming the door shut. Lord Hertford is announced. Hertford tells the "prince" that his father, the king, wants to see him.

Scene Six. Offal Court. John Canty is dragging Edward home, thinking he is Tom. Edward's assertions that he is the prince infuriate Canty, who starts to beat him. Father Andrew steps in to stop the beating. But Canty, in a blind rage, beats the cleric instead, surrounded by the crowd (which hides the actual blows). Canty drags Edward off.

Scene Seven. A hallway in the palace. Everyone is discussing the encounter that just took place between the king and his mad son. It is announced that tonight there will be a pageant along the Thames in honor of Edward's official installation as Prince of Wales.

Scene Eight. Inside the Canty rooms. Edward meets Tom's mother Alice and his sister Nan. Edward insists that he is the prince, and the women plead for Canty to have mercy on the "mad" boy. Canty relents, but threatens them all if they don't make enough money begging tomorrow; he reminds them what it takes to survive. He douses the lights and orders everyone to go to sleep.

Scene Nine. The prince's chambers. The Earl of Hertford is discussing the situation with Lord St. John. Although King Henry is dying, he has insisted that the Duke of Norfolk be executed. Henry is unable to sign the decree himself; the Lord Chancellor can, but only with the Great Seal. The Seal was last seen with the prince, who because of his apparent madness can not recall its whereabouts. When Tom hears of Norfolk's impending execution he asks if it can be averted, but Hertford says that the king is adamant. St. John leaves, and Hertford tells Tom that King Henry wishes the prince to conceal his "infirmity" and stop denying that he is the Prince of Wales. Hertford believes that the boy's memory is faulty and suggests that, with the installation pageant pending, it would be best to let Hertford instruct him in appropriate behavior.

Scene Ten. The Canty rooms, and then the streets outside. All are still asleep except Alice Canty; she is worried about her son's madness and troubled by something else as well, an indefinable strangeness about him. She decides to test him by making a sudden loud noise and shining a light in his eyes; if he responds as Tom would, then he is her son. He does not, and Alice begins to suspect that he has been telling the truth. With sounds of celebration coming from the street, a neighbor knocks at the door and tells Canty that Father Andrew has died from his beating. Canty decides that the family must leave immediately to avoid the authorities; if they are separated, they are to meet him at London Bridge. They leave quickly, Canty dragging Edward. But as they reach the streets, filled with people celebrating the installation of the Prince of Wales, Edward manages to pull loose and run off. Canty angrily follows him.

Scene Eleven. Outside the Guildhall. A wild celebration is going on in honor of the installation of the Prince of Wales, which is taking place within. Edward has managed to make his way here without being caught by Canty. He demands entrance to no avail, and he is taunted and threatened by the crowd. Miles Hendon, a handsome nobleman down on his luck, steps in to aid the youngster. Canty enters and tries to grab Edward but backs off when Miles threatens him. Miles takes Edward away, but Canty follows them at a distance.

Scene Twelve. Miles's room on London Bridge. Miles and Edward converse over dinner. Miles explains that he served in the continental wars for three years, then was captured and spent seven years in an enemy prison. He has just returned to England. Edward tells his story; Miles thinks the boy is mad, but kindheartedly shows him the deference he demands. Edward falls asleep, and Miles decides to bring the boy to Hendon Hall, his ancestral home in Kent, where his family will welcome them. He takes Edward's measurements with a piece of string and leaves to get Edward some decent clothes. Edward wakes from a nightmare, thinking he is back in the palace, but realizes that this nightmare is real. A minute later there is a knock on the door; Hugo, a ruffian in his late teens, tells Edward that Miles is hurt and needs him to come to the south end of London Bridge immediately. Edward follows him off.

Scene Thirteen. Inside the Guildhall. Tom is eating a sumptuous meal, sitting at a table surrounded by lords and ladies, with the assistance of many servants. He has not completely unlearned his old eating habits—he blithely drinks from the fingerbowl—but everyone treats his behavior as normal. A messenger enters and announces that King Henry is dead. All salute Tom as the new king. He immediately pardons the Duke of Norfolk.

Scene Fourteen. Miles's room on London Bridge, and outside. Miles returns to the inn with a set of used clothes and finds Edward missing. As a servant enters with breakfast, Miles questions him and learns that while he was out, Edward was given a message, purportedly from Miles, to meet him at the southern end of the bridge; when Edward left the inn he was followed by a man who, from his description, sounds like John Canty. Miles rushes out of the inn to find Edward, while we hear the news of King Henry's death spreading through the crowd; Edward is the new King of England. End of Act One.

(*Since we will not be taking our adaptation further, we summarize Act Two very briefly*).

Act Two. The second act alternates between scenes featuring Tom—who, with the aid of Hertford and Princess Elizabeth, becomes steadily more comfortable with his role as the king—and scenes featuring Edward and Miles, each trying to reclaim his rightful position amid many perils and setbacks. Just as Edward's role has been inadvertently usurped by Tom, Miles learns that his younger brother Hugh has appropriated his position and his fiancée. Despite many harrowing adventures, in the nick of time Edward manages to prevent Tom from being crowned in his place, and once he is king again, he sets all aright. Tom's father has disappeared, perhaps thrown in jail or murdered by his cronies, and Tom is reunited with his mother and sister.

As you can see, even in the first act we have made a number of changes for dramatic purposes. Let's examine some of them.

The first two chapters of the novel merely describe the births of the two boys and their present circumstances. There is no reason to include either in a stage adaptation. Instead, we can fold all the necessary exposition from the first two chapters about the time, the place, Tom, the Canty family, Father Andrew, Prince Edward, and the vast gulf between rich and poor into a single scene in the streets of London.

Most of the next few scenes follow the action of the novel closely, with small adjustments. For example, instead of having Tom just "find himself" at the royal palace (as in chapter III of the novel), we can increase the dramatic momentum if the appearance of the prince is what stimulates Tom to follow him and attempt to approach him (in Scene One of the musical). Similarly, the plot moves faster if Canty sees the prince and mistakes him for Tom immediately after he has been ejected from the palace (Scene Four), rather than after a gap of several hours (chapter IV). The beating by the Christ's Church orphans in chapter IV has a small "payoff" at the very end of the novel, but this would be lost in a stage version, and would require adding a number of children to the cast for a single scene, so we cut it.

Chapters V through IX of the novel are spent with Tom in the palace. We decided that to alternate between scenes of Tom and of Edward would provide more variety and strengthen the momentum of both plot lines. Except for Scenes Eleven and Twelve, the remainder of Act One alternates between them.

A crucial element in the plot of the novel is the question of the whereabouts of the king's Great Seal. This ultimately pays off in two ways: It allows Edward to prove that he is indeed the king—if the reader accepts the absurd notion that there is no other way to prove that he, and not a beggar boy, is Edward Tudor—and it allows Twain to finish the coronation scene with a laugh, when Tom reveals that he has been using the Seal as a nutcracker since Edward left the palace. But even if we keep both aspects in our adaptation, there is no need to give the discussion of the fate of the Duke of Norfolk as much time in the musical as it takes in the novel. It can be mentioned briefly in Scene Nine. This also allows us to bypass the awkward question of why Hertford never describes the Seal to Tom, even though he knows that the "prince's" memory is faulty.

We have eliminated Tom's first royal dinner (chapter VII in the novel); while it has great potential for comedy, it does little to move the story forward. Instead we have taken some of it and moved it forward to the celebration dinner in the Guildhall (Scene Thirteen). There is a danger in this transfer, of course, because the celebration dinner takes place later in the story, and Tom's gaffes would be more serious at this important public occasion than during his private dinner. But we think it is possible to walk the fine line that allows him to remain a little gauche (and thus funny) but no longer utterly crude (and thus shocking to the lords and ladies). We have also eliminated the river pageant; although it is interesting to read about and could be quite a spectacle, it would take an enormous amount of money and stage time and contribute nothing to the story's momentum.

We have moved Canty's encounter with Miles and Edward, placing it outside the Guildhall (Scene Eleven) instead of outside Miles's lodgings on London Bridge. This eliminates an extra scene we would have had to add before Scene Twelve; although it would be easy enough to do, it would add stage time and possibly an extra set change. Also, if we show Canty following Miles and Edward it makes him seem more sinister and menacing. Scene Twelve corresponds to chapter XII of the original, but—following the golden rule of "show, don't tell"—we have dramatized the scene of Edward's being lured away with a phony message. This also allows us to introduce Hugo earlier and expand his part a bit. By the same golden rule, however, our cutting the scene between Henry VIII and Tom (from chapter V in the novel) and putting it offstage obliges us to replace it in Scene Seven with an entertaining musical number.

In the novel Tom and Edward both learn of the death of Henry VIII in chapter XI. Then follow the scenes in which Miles and Edward become acquainted, Edward is lured away, and Miles goes after him. We have shifted things so that Miles and Edward get acquainted first, then Tom (as well as the audience) learns of the king's death. Miles, along with the rest of London, only hears of it in the next scene, as he sets out in search of Edward. Thus we combine the suspense of Edward's disappearance with the irony that the homeless and now missing boy has become King of England, all of which adds to this scene's effectiveness as the end of Act One.

If we continued our scenario all the way through Act Two, we would make more substantial changes. For example, much of the second half of the novel is taken up with the misadventures of Edward and Miles, and Tom almost disappears from the story; he does not appear at all between chapters XVI and XXX. While the Miles-Edward plot line unquestionably has most of the suspense in the novel, we think that on stage such an imbalance would bother the audience, if only subliminally. So between the Edward-Miles scenes we would occasionally intersperse scenes to show Tom learning more about court customs and manners from Edward's sympathetic half sister Elizabeth and from Lord Hertford. Even as he becomes more comfortable with his role as the new king, the fear could persist—exacerbated by his growing desire to remain in his new role—that he may yet make some gross error that could be his last.

Giving Tom more to do in the second act would not be simply for the sake of variety. To see Tom becoming more comfortable with his royal role might increase the apprehension—even in those audience members who know the original story—that in this version Tom may decide not to give up his power to Edward. It would also allow us to build up to the scene of the coronation procession—when Tom, now ready to become King of England, is confronted by his mother and denies her—more subtly and believably than in the novel. This moment, which is almost thrown away in the novel, could be one of the most powerful scenes in the musical, and giving Tom's character more of an arc toward this moment would also make explicit what is only implicit in the novel: that his remorse following the rejection of his mother motivates Tom to readily acknowledge Edward as the true king, when he could easily have had Edward thrown in jail as an impostor and become king himself.

Just as we would augment Tom's role in the second act, we would eliminate some of the Edward-Miles material. The portion of the novel after Edward's kidnapping takes two-thirds of the entire story, and almost all of it is about Edward and Miles. We would pare this down, not only to make room for the scenes with Tom, but also because much of it consists of melodramatic incidents that trace the same emotional paths over and over. Certainly the entire section dealing with the insane hermit could be cut without losing anything significant. We might condense the growing antipathy between Hugo and the prince into two scenes, the second of which would show Hugo framing Edward for theft and having him arrested. More likely we would cut that incident entirely—since it would involve additional sets, a large crowd, and at least two or three additional scenes—and simply have Miles catch up with Edward as he escapes from Hugo after the failed con game. This would then lead us directly to Hendon Hall, where Miles's misfortunes would parallel and intensify those of Edward. We would probably keep the whipping scene—although we might move it offstage, since the live spectacle might be upsetting to children. But we would eliminate Miles's sitting in the pillory, to save time. And of course both the climax, which in the novel has Lord St. John running back and forth between

Westminster Abbey and the royal palace several times in search of the Great Seal, and also the denouement would need to be compressed and made more theatrically effective. (The 1937 movie is almost ruined by this repetitive tedium so near to the end.)

We would probably also add at least one scene in Act Two for Alice Canty and her daughter Nan. In the novel, Tom's mother does not appear again between the hurried departure from Offal Court (in chapter X) and the procession where she recognizes her son (in chapter XXXI). His sisters do not reappear at all, nor are they even mentioned. A scene with Alice and Nan seems important to explain what happened to them, as well as to show the missing stage in Alice's character arc between her suspicion that Edward is not her son and her sudden appearance at the procession. Rather than this appearance being sheer and rather farfetched happenstance, as it is in the novel, we could show Alice actively searching for her son and thus being more likely to recognize him, even in the midst of pomp and ceremony. This also would help to build up a character whom we see as important.

Sample Project B: An Original

THE STORY

As we have indicated, originals are hard to write, which explains why there have been comparatively few of them. But precisely for that reason, we felt it was important for this book to include a sample original project.

So we did exactly what we would do if we were a writing team looking for an idea: We sat in a room and started to brainstorm. One of us had an idea for a story about the phenomenon of so-called unidentified flying objects (UFOs), and the places in the American Southwest where extraterrestrial vessels have supposedly appeared or landed. The other author suggested a story about children whose parents are divorced, leaving their lives and emotions split between two people and two households. This is a common situation in our society, and it usually burdens the children with pain and emotional conflicts that they may never resolve.

It seemed clear that neither idea was enough for a good musical. The UFO idea had a number of positive features: It was distinctive, it suggested a large canvas of events, and it had never been utilized in a successful musical. On the negative side, it had little inherent emotion, and the plots it suggested all seemed either melodramatic or overly reverential, reminiscent of movies like *ET, Starman, The Day the Earth Stood Still,* and *Close Encounters of the Third Kind.* The children-of-divorce idea had different positives: a real situation to which almost everyone could relate directly or indirectly, and plenty of potential emotion. On the other hand, it was extremely general and had no larger-than-life quality, nor anything more distinctive than many made-for-television movies.

We decided to combine the two ideas. The combination seemed to give us the best of both worlds: the interesting milieu of the Southwest and UFO lore, and the strong emotional potential and universality inherent in the situation of children with divorced parents. Our working title: *Life on Another Planet*.

Since we had already picked *The Prince and the Pauper* for our other project, we decided immediately that this project would be aimed at a slightly different audience; it would be a show for adults rather than another family show, although it might well appeal to older children. For the same reason, we were reluctant to have a young boy as the protagonist of our original. We decided to make the character a girl rather than a boy, and more important, an adolescent rather than a child. This had the additional advantage that professional actor-singer-dancers who could pass for fifteen years old would be much easier to find than those who could pass for ten. We also decided that her parents' conflicts and emotions could be just as interesting and affecting as hers. So she would not be *the* protagonist, but only one of three. This would also make casting easier, because the actress playing the daughter would not be carrying the entire show on her shoulders.

At first we planned to set the show in New Mexico. The mother, a sculptor, could live in Santa Fe, which has a thriving arts community, and the father could be a geologist living in Carlsbad, doing research work at Carlsbad Caverns. These locations gave us two interesting and contrasting milieus for the parents. Furthermore, when the daughter shuttled between them, she would pass close to Roswell, the town where the first modern sightings of UFOs were reported in 1947. This made it natural that our teenage girl, whom we had already decided must be intelligent and curious, might speculate and fantasize about alien beings—especially if her own situation was an unhappy one.

However, after a good deal of thought and research, we decided to shift the location from New Mexico to Nevada. There were several reasons, most of them logistical ones concerning the plausibility and likelihood of car or bus trips between Carlsbad and Santa Fe. So instead of a geologist who lives and works in Carlsbad, the father has become a desert ecologist who lives near Las Vegas and works at the Nevada Desert Research Center (an affiliate of the University of Nevada at Las Vegas); part of his research takes place in an area granted to the Desert Research Center within the Nevada Test Site, the huge U.S. military area northwest of Las Vegas. (The Test Site is where the government used to conduct above-ground nuclear bomb tests.)

The mother, still a sculptor, now lives at Lake Tahoe. The drive between Tahoe and Vegas goes past the Test Site, and in particular the mysterious Area 51, which, both in reality and in our show, is a secret government installation within the Nevada Test Site and a part of UFO legend. While the U.S. government makes no public acknowledgment of Area 51, aircraft are not allowed to fly over it, and it is common knowledge that top-secret military aircraft and weapons are built and tested there. Some people, however, believe that it is an area where alien spacecraft have appeared and landed, and that it is kept secret because the government does not want the public to learn the truth about UFOs. In fact, a section of highway that passes by the Test Site grounds has been dubbed the "Extraterrestrial Highway" by locals, and sports a number of tourist attractions related to UFOs.

All of this gave us a situation and a milieu, but it didn't give us a plot, although we knew that somehow the daughter would eventually find herself in or near Area 51. However, one of us knew a boy with a situation similar to that of the daughter; the boy spent most of the year living with his mother and her new husband, and he flew across the country to join his father for holidays and summers. The parents agreed that at the age of thirteen he could decide which parent he would live with until college. This suggested a crucial plot point for our musical.

We decided that we did not want the daughter to bring her divorced parents back together at the end, as in *The Parent Trap* and many similar movies, although she might fantasize about it. Such an ending felt false and too "cute" for an adult musical. So there must be a different resolution, and she must come to accept that her parents would not reunite and bring back the nuclear family of her childhood. We felt that this was better than a "parent trap" resolution, because it would be both more realistic and more emotionally complex than a conventional "happy ending."

Again we decided that this would be a traditional book show with dialogue. Obviously the speech would have to be contemporary, although we would avoid using actual teen slang because it dates so quickly. The music would also need to have a contemporary sound, but that would not require any particular current style.

THE CHARACTERS

Like other aspects of original musicals, the characters have to be created from scratch. But as we continued to brainstorm about our story idea and what it needed to come to life on stage, certain characters started to become clear to us.

Initially we vaguely assumed that the daughter would be a "typical teenager"—not that there has ever been such a thing. But in integrating our two basic ideas it became clear that for aliens and UFOs to matter to her, even in idle fantasies, she would need to be bright, interested in science, and something of a loner. On the other hand, we didn't want her to be seriously delusional about alien visitors; that would be another show entirely, and probably not a good musical! So our first character description was as follows:

Kari Auster. She has just turned fifteen. She is very bright, unconventional, a loner. She is not a "joiner" and has only one real friend at Lake Tahoe, where she has been living. She loves science, and is especially interested in astronomy and xenobiology. She has a decent-sized telescope, a gift from her parents, which she brings with her each time she shuttles between them. She is also curious about UFOs. She fantasizes about being the first person to discover incontrovertible proof of alien life, or the first to make definite alien contact; sometimes, when depressed about her parents, she fantasizes about belonging with aliens, being an alien changeling, or being taken away by aliens.

Next we had the parents, whose characters demanded a delicate balance. We wanted there to be a real tug-of-war in the daughter's heart between her parents, and we didn't want the audience to feel that we had taken sides. Therefore both parents needed to be essentially good and loving people, neither abusive nor cold. They needed to be different in ways that would explain why they split up; yet they also had to have things in common, or it wouldn't be plausible that they had fallen in love in the first place.

We considered it most likely that the mother would have primary custody of the child. But we also felt that one parent should be in a new relationship that was stable and going well. If this were the father, then the mother would be at a severe financial and emotional disadvantage; most women still earn less than men, and divorced women are less likely to remarry than divorced men. So we decided to make the father single and give the mother the relationship. But since Kari would probably already be living with her mother, we needed to balance matters a little more, so that there would be a fifty-fifty chance that she might now choose to live with her father. We decided to reverse the gender stereotypes: Both of her parents love her, but it is her father David, although alone and very busy with his work, who gives more time and attention to Kari. On the other hand Terri, the mother, has a whole new group of friends in a new town, and a new boyfriend with his own family, which intensifies the contrast between the parents' living situations. Terri is also very excited about the recent flourishing of her career as a sculptor—which to some extent she had put on the back burner while married to David and raising Kari. So although she is a charming and vivacious woman, with all these new concerns she never seems to have time to really focus on Kari. Many of the details in David's and Terri's profiles stemmed from these considerations, and we added other details from our own experiences and observations. Our characters now included:

David Auster. David, about forty, lives by himself in a small house in the suburbs of Las Vegas. He is a desert ecologist who works at the Nevada Desert Research Center and teaches at the University of Nevada at Las Vegas. He is quiet, introverted, serious, organized, and devoted to his work, but also warm and nurturing. He knows almost everything going on in Kari's life—unlike Terri, with whom Kari lives for most of the year. David is very busy, though, and doesn't have much time to spend with Kari even when she is staying with him. Since his divorce he has had several girlfriends, but there is no one in his life right now. He can be very funny, in a wry quiet way. He dislikes Las Vegas but loves the desert and its solitude. He spends a lot of time there for both work and pleasure; he has taken Kari on rafting and hiking trips through the Grand Canyon and other parts of the Southwest. Kari loves their time together, but when David is working she is all alone, especially during the summers. She has no friends in Las Vegas and dislikes the city. Also, although she respects and loves David, Kari is somewhat embarrassed that he is such a loner and a "science geek," perhaps without realizing that she is very much like him.

Terri Auster. Terri is about thirty-eight. She lives in a condo at Lake Tahoe. She is a sculptor and has recently begun to make a good living from the sale of her artwork. She is attractive and youthful, disorganized, flamboyant, gregarious, fun-loving, cheerful, and loved by everyone who knows her. She likes crowds and ferments of artistic and intellectual activity. She is also very involved with her career, her new community of friends, and her boyfriend Hal Ferris. Kari loves her mother, but also resents that Terri never seems to have enough time for or interest in her.

David and Terri met in graduate school at the University of California at Los Angeles, sixteen years ago. David, an ecology and biogeography major getting his doctorate, was from Tucson. Terri, an art major going for her master's degree, was from Berkeley. They met on a sunny afternoon, hiking in Topanga Canyon. Although quite different in backgrounds, habits, likes, and dislikes, they had a number of things in common. Both had a good sense of humor; both loved foreign languages and other verbal challenges such as cryptograms and crossword puzzles. Perhaps most important was that both loved the desert and outdoor activities like rafting and hiking. They were also strongly attracted to each other, even though he tended toward the "geeky" and she toward the "flaky."

They married a few months later, the summer before their final year at UCLA. Kari was born a year later. At the time, Terri assumed that they would remain in California near San Francisco or Los Angeles, but David got a job with the Nevada Desert Research Center and a chance to combine that with teaching at the University of Nevada at Las Vegas. They moved to Las Vegas and lived together for eight years, till Kari was seven. During this time their differences started to drive them apart. Terri found David too obsessed with his work, too neat and organized; David found Terri too loud, too social, more interested in going out to noisy places and meeting other people than in staying home or going out with just him (and Kari). Also, David would have liked to have at least one more child, but Terri felt that she had been neglecting her own career, and resented the expectation that she would postpone it indefinitely until their children were grown.

Their divorce was uncontested and, compared to many others, essentially amicable—at least on the surface. Shortly afterward, Terri moved to Lake Tahoe. They agreed that Kari would live with Terri during the school year, and with David for most of each summer, until she reached high school. Since then—for eight years, or more than half her life—Kari has shuttled back and forth between them. (Most of the time David has driven Kari both ways, because he is more willing to do it, and because Terri is not a good or patient driver.)

During this time, relations between David and Terri have been volatile and uneasy. At first everything was as amicable as they could make it. Then, over the next few years, as they had to deal with practical questions of raising and sharing Kari, a great deal of anger and resentment emerged on both sides, and for a couple of years they barely spoke to each other. Since then, things have gradually mellowed, but both the caring and the resentment linger; relations between them are generally calm and pleasant, but certain things can trigger violent arguments. This bothers both of them because they prefer not to lose their tempers and feel that they almost never do—except when dealing with "the ex." In addition, now that Terri is in a stable relationship and David is not, David feels at a disadvantage, although he knows this is irrational. He also worries about how the changes in Terri's life are affecting Kari. On the other hand, Terri is afraid that David's continual solitude, both personal and professional, has made him withdrawn and strange. She also knows that when Kari stays with him he is often away working, and she worries about what Kari might drift into, out of boredom.

Once we had roughed out our three central characters, we needed to fill in the secondary characters that would be essential to our story. These included:

Hal Ferris. Hal, fifty-two years old, is a widower who lives at Lake Tahoe. Handsome, tanned, in good shape, he has a flourishing business in construction and building materials. He is originally from Reno and owns stores there, in Carson City, and at Tahoe. He has two children, Carolyn and Mike. Hal has a pleasant and engaging personality, and tries not to judge people. He adores Terri and indulges her; he likes her kookiness and "artsiness," although he doesn't share it. Terri loves that he shares some of her interests and activities, that he is gregarious like her, and that he is very supportive of her career. They met at his Tahoe store when it opened last year; she went there to buy materials for her sculpture. While well off, he is not rich. He loves outdoor activities like tennis and water sports like sailing. But he doesn't like hiking much, and doesn't care about the desert at all. David realizes that it is Terri's situation he resents and not Hal personally, and because Hal is very pleasant and easygoing, David gets along with him better than with Terri.

Mike Ferris. Mike, Hal's younger child, is eighteen and about to enter his senior year of high school. He is big and beefy, a football player, and an average student. Like his father, he takes people and events as they come. He is friendly with Kari, but doesn't feel particularly close to her; they find little to talk about. Mike is interested in football, cars, and hanging out with his jock friends. He hopes to get a football scholarship to a Big Ten college.

Carolyn Ferris. She is twenty-two, athletic and attractive, but not particularly intellectual. She graduated from college last year and recently married Brad Tyler. They now live in Phoenix.

Brad Tyler. Brad is twenty-five. He is handsome, athletic and outgoing. Brad is a computer systems salesman, so charming and successful that his company made him the sales manager of their Phoenix office. He grew up in Reno near the Ferrises and gets along well with Hal, as he does with everyone else. He and Carolyn get together with Hal and Mike, and now Terri, several times each year.

Jen Haber. She is fifteen years old and lives at Lake Tahoe. She is Kari's best friend and in her class at school. Like Kari, she is very bright and tends to be a loner. She also comes from divorced parents. Unlike Kari, she deliberately affects a flamboyant rebelliousness, which manifests itself in frequent changes of hair color and various pierced body parts. She wants to be a filmmaker or performance artist and sees herself as a misunderstood genius. Although she and Kari are very close, she does not see Kari during the summer months except for occasional visits, and she does not share Kari's curiosity about aliens, except as potential subject matter for her art.

Zirr. Zirr is a mysterious individual, tall and androgynous in appearance; it is impossible to tell if Zirr is male or female. (Zirr might be played by a woman with a low voice.) Zirr also speaks English somewhat strangely, with an unusual accent, and seems strange enough to be an alien, if aliens resemble human beings.

Tertiary characters such as Terri's artistic friends and Mike's jock friends would also appear in Act One; others would probably come up during the writing of Act Two. At this point we did not envision an ensemble per se, but rather a group of actor-singers who would play a number of small roles such as party guests, people on the streets of Las Vegas and other Nevada towns, tourists along the Extraterrestrial Highway, state police officers, government security agents, and perhaps an extraterrestrial or two.

THE LIBRETTO

Since *Life on Another Planet* is an original project, and there is no source material with which to compare it, we will simply present the scenario we have developed. As you will see, Act One is reasonably complete, while Act Two is only briefly sketched.

LIFE ON ANOTHER PLANET
A Musical in Two Acts

Act One. Scene One. It is mid-May, in a big chalet-style house overlooking Lake Tahoe, Nevada. Kari Auster, who has just finished ninth grade, is at her fifteenth birthday party. Her mother Terri and her father David are both there, but we soon learn that they are divorced; that David lives outside of Las Vegas, five hours away; and that the house belongs to Terri's boyfriend, Hal. Terri still has her own apartment, but she spends most of her time here, and therefore Kari does too. Both of Hal's children, Carolyn and Mike, are there, as well as Carolyn's husband Brad, friends and business associates of Hal's, and artistic friends of Terri's. There are few children of Kari's age there other than her best friend Jen. For seven years Kari has been living with her mother at Lake Tahoe, and spending holidays, spring breaks, and summers with her father in Las Vegas. But according to the divorce settlement between David and Terri, by August 1st Kari must decide whether she will continue to live with her mother at Tahoe, or move to Las Vegas to live with her father for the next three years. She is already feeling pressure about this decision. The usual arrangement is that she spends most of the summer with David, except for the July 4th weekend and the two weeks afterward. But she asks her parents if this summer she can spend two weeks at a time with each of them, to help her make up her mind. Thus she will spend the next two weeks with David, go back to Terri in mid-June, and so on, until August 1st. She feels utterly confused about what to do. A birthday cake is brought in and she tries to blow out all fifteen candles, but she can't. She wonders if that's because she isn't sure what to wish for.

Scene Two. The Nevada desert, about halfway between Lake Tahoe and Las Vegas, along the "Extraterrestrial Highway." The next day. David and Kari have been driving south, and have just left David's car parked by the side of the road.

Kari has gotten her learner's driving permit, and they have stopped for a break, after which Kari will drive for a while. They talk briefly about the summer, but to change the subject Kari starts discussing her interest in Area 51, which is not far away. David notices some unusual plants nearby and leads Kari over to examine them. We see his, and Kari's, love of the desert and of strange and exotic forms of life. David tells Kari how Native Americans used to go out into the desert alone to seek a vision, for insight and wisdom in how to live. He talks about the "web of life," the interrelatedness of all things, by which he means life on earth—but which for her includes all life in the universe.

Scene Three. At David's house in Las Vegas, and along the Las Vegas Strip. This is a montage of short scenes over the next two weeks. Kari is staying with David. She is happy to be with him there, or to go with him on field trips into the desert, but when he goes to work she gets lonely. At night she can go into the desert and watch the stars through her telescope, but she spends many days either watching "infomercials" on television or wandering the streets. She hates Las Vegas and feels strongly that she does not fit in. It's hard for her to imagine living here for the next three years.

Scene Four. Inside the Flying Saucer Rest Stop, June 1st. This is a diner, souvenir shop, and motel in Dry Lake, a small town about halfway along the Extraterrestrial Highway. David and Kari are stopping for lunch on their way back to Tahoe, as they have many times before. It is a colorful place; there is a big sign saying "Aliens Welcome" at the entrance, and photos of UFOs and other strange phenomena cover the walls, along with eight-by-ten photographs of famous movie monsters. The souvenir shop is at the far end of the dining room, which is less than half full. David offers to help Kari decide where to live by employing logic and a list of pros and cons, but she does not want to discuss it, and soon jumps up to look at the souvenir shop. Zirr, an androgynous individual with strange mannerisms who is browsing the book racks, notices her and observes her closely in a way that might be considered spying. Soon Kari gets bored. She and David leave, as Zirr pays for a pile of books and magazines. Looking out the window after them, Zirr speaks in a strange language into some sort of cellular telephone.

Scene Five. Hal's house at Lake Tahoe. This is a montage of short scenes over the next two weeks. Still trying to make up her mind, Kari is afraid that she does not belong here either. Hal is all right but he's not her father, and she can't really talk to him—or to Terri—about her interests; there are always too many people around, including Mike and his jock friends, Hal's suburban bourgeois friends, and Terri's art gallery friends, whom Kari finds pretentious and phony. She likes Tahoe but misses the desert; she likes Jen, but few of the other kids at school. She wonders if she'll ever have a boyfriend. Terri is sympathetic but, as always, the center of a whirl of activity: sculpting, playing tennis, entertaining, networking, spending time with Hal. Kari wonders if her mother really wants her as much as her father does.

Scene Six. Terri's apartment at Lake Tahoe. June 14th, the day before Kari returns to Las Vegas. Kari and Jen are hanging out and talking, primarily about where Kari is going to live. Then Jen has to go home, and as she leaves, Terri enters with big news: She is going to marry Hal in July. She tells Kari that Hal first suggested the idea after his daughter Carolyn's wedding a few months ago; Terri was reluctant at first, relishing her autonomy after the frustrations of her marriage to David, but she knows that this marriage will be good for her, and hopes that Kari is OK with it. She has essentially been living with Hal for almost a year, and using her own apartment mainly for her sculpting. But Hal is going to convert a shed on his property into an art studio for Terri, so she has decided to give up the apartment. Although this is not a complete surprise, Kari is hit hard. She is not sure how she feels about the marriage, but she hates the thought of losing the apartment, which has been an occasional refuge, a quiet place to study or hang out with Jen. She asks Terri why she has to marry Hal, especially right now—why they can't just keep living together. Even though Kari likes Hal, this would be the final nail in the coffin of her parents' marriage. She doesn't like things as they are, but she's afraid that with the wedding and its consequent changes, her life will become even more unhappy.

Scene Seven. Hal's house, the afternoon of the next day. David has come to pick up Kari, who's not home yet. Hal is at work, but Terri tells David about the wedding. He accepts it but still it bothers him, although he doesn't admit it to her. They start talking about Kari's summer and her future, which brings up the past. They start reminiscing, but end up as usual in an argument. Kari comes home in time to hear the end of the argument. David says that there's a lot more to talk about, and insists that when it's time for Kari to return in two weeks, Terri must meet him halfway, in Dry Lake.

Scene Eight. In the car on the way to Las Vegas, later that day. David asks Kari if she has decided whether she wants to live with him or her mother. Still upset by the argument between her parents, she says she doesn't know; what she'd really like is to live with both of them, together. He says that that is never going to happen, and she gets very upset, wildly spinning a fantastic picture of the three of them living together in harmony with nature and each other, halfway between Tahoe and Vegas, finding all they need in the desert, where they can sculpt, research, and look for alien spaceships.

Scene Nine. David's house in Vegas. Early morning on July 4th, the day Kari is to return to Tahoe. Kari comes out of her room as David is leaving for his daily run. When he is gone, she brings out a big new hiking backpack, which she starts to pack with outdoor gear and food. She calls Jen at Lake Tahoe and says her parents are still fighting about her, and she can't fit in at either place. She tells Jen that she is running away; she doesn't know where she'll end up, but right now she's going into the desert to seek insight and wisdom. She needs time and solitude to figure out where she belongs. She can't live with either parent; neither one really understands her, and she doesn't feel at home in either Tahoe or Vegas. She does not pack her telescope.

Scene Ten. A street in the center of Dry Lake, later that day. David and Kari are sitting on a bench, waiting for Terri, who is late as usual. Kari tries to convince David that it's all right for him to leave without waiting for Terri to show up, but he refuses to leave her there alone. She tries to pick a fight with him and, when that doesn't work, sends him down the street to get her something to eat. As soon as he is gone, she hoists her backpack and heads out into the desert. End of Act One.

As with *The Prince and the Pauper*, we have not constructed a detailed scenario for Act Two of *Life on Another Planet*. Here is the first scene, followed by a brief summary of the rest of the show.

Act Two. Scene One. The Dry Lake street, a few minutes later. Terri is sitting there when David returns. They assume that Kari has gone to the bathroom or into a store and will be right back. They start discussing the wedding and get into another argument, in the middle of which they realize that Kari has been gone a long time. They disagree on how to search for her, and separate.

The action of Act Two has two plot lines. In one, David and Terri search for Kari, separately at first. When that doesn't work, they reluctantly decide to join forces. Working together to find Kari, while it does not magically make everything all right between them, helps both of them to put the bad feelings of the past behind them and to be friends. In the other plot line, Kari finds herself in the desert near Area 51. At first confident that she can survive in the desert, she soon finds it tougher than she thought. In a bad fall she twists her ankle. She is getting worried and frightened when along comes Zirr, as though on a stroll. Zirr bandages her ankle and calms her. In answer to her questions, Zirr gives the impression of being someone who works in Area 51—perhaps a foreign technician or scientist—and likes to hike through the desert. But Zirr looks and acts so strangely, and gives such elliptical answers to Kari's questions, that Kari suspects that Zirr is an extraterrestrial. She wonders if Zirr lives in Area 51 as a guest of the government, or perhaps Zirr is an alien who has secretly landed in the desert in an attempt to contact fellow aliens held captive in Area 51.

Zirr's utterances are cryptic and strange, but they seem full of wisdom to Kari. As a result of her time with Zirr she decides to return to her parents, and shortly afterward, with the help of rangers, police, and security men, they find her. Kari now accepts that David and Terri are better off apart; and that when Terri remarries, Kari won't lose either parent, but rather will become part of a larger family. The show ends without Kari having chosen where to live; now that Terri and David can get along with each other, she sees that her decision is no longer as crucial as she had thought, because either choice would probably work out fine.

You have probably noticed that, both before and within the scenario of this show, we have included considerable detail about the characters and their "backstories." This is especially important when all the characters are created from scratch.

We have mentioned that the scenario of Act One is rather detailed, while that of Act Two is very skimpy. As with *The Prince and the Pauper*, it has never been our intention to finish writing this show. Because of that, we have not worked out the plot of Act Two in detail, although we do have a general idea of its course and resolution, and a number of specific ideas as well. Obviously the question of Zirr's identity would have to be answered. Our own feeling is that Zirr would turn out to be not an extraterrestrial but only an eccentric who works in Area 51 or lives nearby. If we were writing the entire show we would almost certainly reveal that. We might exercise the option of leaving the question open at the end—but that might seem like a "cop-out," or too reminiscent of an episode of *The Twilight Zone*.

The plot line of David and Terri's search for Kari would undoubtedly include scenes with Hal Ferris and Jen Haber, and probably Mike Ferris and Brad and Carolyn Tyler as well. Ideally some of these characters would be part of a subplot that would complement the main plot thematically.

Once the scenario is done the writers can go on to the next stage, the spotting of the score, which is best done with all collaborators present and contributing. Then the librettist can start to flesh out the scenes with details and dialogue, and the songwriters can start to choose and write their songs. Since neither of our projects is an actual show, however, we will not work out the librettos with much more detail than we already have, except for the scenes containing the songs that we will write in the next chapter. If we were writing an entire musical, of course, we would not go on to the next stage without having worked out the scenario in detail all the way to the end.

Theoretically, one could write a show from one scene to the next without even a plot outline. Some playwrights claim that they start writing a play with no idea where it's going—or even what it's about—and they go where their characters lead them, as though they were simply taking dictation. We have never heard of a successful musical being written this way, but it is possible.

Many of the skills necessary to write a play can be learned from books. (Some of the best are listed in Appendix B.) Beyond that, chapter 3 covers most of the important principles of libretto construction. There are, however, two other things we consider necessary for anyone planning to write a libretto. The first is careful study of the librettos of the classic musicals, a list of which can be found in Appendix D. To try to write a new musical without first being familiar with the best of the old is as shortsighted as attempting to write a tragedy without having read Shakespeare and the Greeks. The other necessity is practice. Musical theater is nothing if not a pragmatic genre. As with learning how to write songs, the only way to truly learn how to write a libretto is to do it—and then do it again and again.

Chapter 8

Writing the Score

Song Spotting

When the primary and secondary characters and the plot scenario have been worked out in detail, the spotting of the score can begin. Songwriters are often so eager to get started on the songs that they don't wait until this stage has been reached. But if you start writing songs at random before you have a relatively firm scenario, there is a good possibility that some or all of those songs will end up having to be either rewritten or cut. Sometimes it's useful to try a song at an early stage, just to get the "feel" of the characters and the show, but in that case you need to keep in mind that your song is an exercise and may not end up in the show.

Song spotting can proceed in a number of ways, but a useful approach is simply to look through the scenario or outline, over and over, until you have identified the important emotional and structural high points. There should be a song at each of these high points, unless you decide that certain moments would be more effective without a song—for instance, if you decide that the best place for a song is shortly *after* a high point, when a character reacts to what has just happened. At this stage you do not have to identify the type of song you want, or even exactly who sings it, unless a choice clearly suggests itself to you. Sometimes waiting to choose is better, because the more familiar with the characters and material you become, the more likely you will be to make the best choice for a particular moment.

Once the essential songs have been identified, you can start picking the places for optional songs. Spotting songs is a little like putting candles on the top of a cake; it is not merely the number that counts, but their placement as well. Optional songs are sometimes chosen for variety. For instance, you might set a comedic moment as a song if you already have a number of serious songs around it, or you might feel the need for a group song if there are many solos. Sometimes optional songs are chosen for practical reasons—for instance, you might need a big number to cover a major scene change, although in recent decades that has rarely been necessary. As another example, if you write a show for a star, you will have to write a number of songs for that star to sing, and you will have to find places for those songs.

Another consideration to keep in mind is balance. In writing *The Prince and the Pauper*, one assumption with which we started is that each of the two leads, Tom and Edward, should have about the same number of songs to sing. This assumption might change, however; the balance of songs between the two boys would depend on a number of factors. For instance, since Edward has most of the difficult dramatic material, we might decide that he could be cast with a boy who acts well but doesn't sing well, and Tom could be cast with one who does sing well. In this case Tom would probably have more songs than Edward. Another option would be to make both boys' parts speaking but not singing roles, and to give all the songs to the adults such as Canty, Mrs. Canty, Miles, and Edith. This doesn't seem the best choice to us, however, because the emotional and structural crux of the story lies with the exchange of roles by the two boys, and it would feel strange indeed to have a musical in which neither of the title characters sing.

Now let's start spotting songs for our two shows. We will reproduce our scene summaries from the previous chapter, but this time we will follow each one with our ideas about song placement.

THE ADAPTATION

The Prince and the Pauper, Act One. "*Scene One.* London, 1547. The streets of London near the inner City. We are introduced to the poor people of Offal Court, including Tom Canty, his family, and Father Andrew. Tom sees Prince Edward and his retinue pass by on the way to the royal palace and follows them."

While an opening number is not necessary, for this show it seems like a natural idea. It would be a big number with almost the entire company—probably an extended musical scene, although we would try to find a song to frame it, as the song "I Hope I Get It" ties together the long opening number of *A Chorus Line*.

This opening would show us the world of sixteenth-century London, introduce many of our main characters, and highlight the vast gulf between the lives of the haves and the have-nots.

"*Scene Two*. Outside the palace. Tom pushes his way through the crowd to catch another glimpse of the prince. Thrust against the gate, he is treated roughly by a guard. From a window the prince sees this, reprimands the guard, and commands that Tom be brought inside."

No obvious song idea jumped out at us in this scene, but if it is short enough it doesn't need a song. Tom could sing a song about the life of a prince, but we think that would be more effectively covered in both the opening number and the song we found in the next scene.

"*Scene Three*. The prince's chambers. The boys compare their lives, and each wonders how it would feel to live like the other. At Edward's suggestion they switch clothes, and they realize that they look alike. Edward notices a bruise on Tom's arm, and rushes out to give the guard another reprimand."

A song idea for this scene emerged immediately. One of the most charming moments in Twain's novel is when Tom describes his life to the prince. This seemed a perfect spot for a duet for the boys, comparing their lives. It could be a charm song, or perhaps a comedy song.

"*Scene Four*. Outside the palace. Edward comes out and starts to chastise the guard, but the guard sees only a ragged beggar boy, and throws him off the palace grounds. Edward protests that he is the prince, but no one believes him. The crowd taunts him and shoves him away from the palace, until he lands in the hands of John Canty, who drags him off."

Nothing jumped out at us here. However, there could be a short song for the crowd mocking Edward.

"*Scene Five*. The prince's chambers. Tom preens in the prince's clothes for a while, but soon begins to wonder what happened to his new royal friend. He looks outside the room, where he is instantly met by armed guards; he panics and jumps back into the room, slamming the door shut. Lord Hertford is announced. Hertford tells the 'prince' that his father, the king, wants to see him."

An obvious idea for a song in this scene is a soliloquy for Tom, before Hertford enters. In the first part of the song Tom would daydream about gifts the prince might bring him. In the last section of the song, after seeing the guards, he would start imagining not gifts but punishments. This could be either a humorous rhythm song or a comedy song.

"*Scene Six*. Offal Court. John Canty is dragging Edward home, thinking he is Tom. Edward's assertions that he is the prince infuriate Canty, who starts to beat him. Father Andrew steps in to stop the beating. But Canty, in a blind rage, beats the cleric instead, surrounded by the crowd (which hides the actual blows). Canty drags Edward off."

Nothing immediately springs to mind for this scene. There could be a rhythm song for Canty or for the scene as a whole—including Father Andrew's intervention and death—but that feels more like a through-composed show than the traditional show we are writing.

"*Scene Seven*. A hallway in the palace. Everyone is discussing the encounter that just took place between the king and his mad son. It is announced that tonight there will be a pageant along the Thames in honor of Edward's official installation as Prince of Wales."

This scene would definitely need a song for the courtiers discussing the prince's mad behavior. Such a song would not only provide another group number, but also fill in the audience about what was said between the king and the pauper. It would also allow us to avoid bringing Henry VIII on stage; to compress the novel's talky scene into a compact lyric; to illuminate the way gossip travels through a group; and to give the perspective of the courtiers, which could prove important at several later places in the show. Because it describes offstage action, it would need to be a very entertaining and theatrical number, probably a comedy song.

"*Scene Eight*. Inside the Canty rooms. Edward meets Tom's mother Alice and sister Nan. Edward insists that he is the prince, and the women plead for Canty to have mercy on the 'mad' boy. Canty relents, but threatens them all if they don't make enough money begging tomorrow; he reminds them what it takes to survive. He douses the lights and orders everyone to go to sleep."

We think there might be a song here for Canty giving his philosophy of life: that it's a brutal world, and that in order to survive you have to be ready for anything and knock down anyone who gets in your way. It would probably be a rhythm song in moderate tempo.

"*Scene Nine*. The prince's chambers. The Earl of Hertford is discussing the situation with Lord St. John. Although King Henry is dying, he has insisted that the Duke of Norfolk be executed. Henry is unable to sign the decree himself; the Lord Chancellor can, but only with the Great Seal. The Seal was last seen with the prince, who because of his apparent madness can not recall its whereabouts. When Tom hears of Norfolk's impending execution he asks if it can be averted, but Hertford says that the king is adamant. St. John leaves, and Hertford tells Tom that King Henry wishes the prince to conceal his 'infirmity' and stop denying that he is the Prince of Wales. Hertford believes that the boy's memory is faulty and suggests that, with the installation pageant pending, it would be best to let Hertford instruct him in appropriate behavior."

Despite the length of the description, this would be a relatively short scene. Still, Hertford's instructions to Tom about how to behave as a prince could make an interesting song, either as a solo or a duet. It could be a charm song or a comedy song.

"*Scene Ten.* The Canty rooms, and then the streets outside. All are still asleep except Alice Canty; she is worried about her son's madness and troubled by something else as well, an indefinable strangeness about him. She decides to test him by making a sudden loud noise and shining a light in his eyes; if he responds as Tom would, then he is her son. He does not, and Alice begins to suspect that he has been telling the truth. With sounds of celebration coming from the street, a neighbor knocks at the door and tells Canty that Father Andrew has died from his beating. Canty decides that the family must leave immediately to avoid the authorities; if they are separated, they are to meet him at London Bridge. They leave quickly, Canty dragging Edward. But as they reach the streets, filled with people celebrating the installation of the Prince of Wales, Edward manages to pull loose and run off. Canty angrily follows him."

Here a song idea occurred to us immediately. This is a perfect spot for Alice to sing a ballad about her concern for her afflicted son. This ballad, which might be the first one in the show, is important for two other reasons. First, it makes Alice's part larger. (In the novel, none of the female characters has any significant lines of dialogue.) Second, following her introduction in the opening number, it comes at a crucial point in Alice's character arc, on its way to intersect Tom's arc during the coronation procession, when he denies knowing her.

"*Scene Eleven.* Outside the Guildhall. A wild celebration is going on in honor of the installation of the Prince of Wales, which is taking place within. Edward has managed to make his way here without being caught by Canty. He demands entrance to no avail, and he is taunted and threatened by the crowd. Miles Hendon, a handsome nobleman down on his luck, steps in to aid the youngster. Canty enters and tries to grab Edward but backs off when Miles threatens him. Miles takes Edward away, but Canty follows them at a distance."

No idea came to mind immediately here, but certainly Miles could have a song when he appears and defends Edward against the crowd; it could tell the audience much about his character and personality. Since we need the audience to root for him, it would probably be a charm song.

"*Scene Twelve.* Miles's room on London Bridge. Miles and Edward converse over dinner. Miles explains that he served in the continental wars for three years, then was captured and spent seven years in an enemy prison. He has just returned to England. Edward tells his story; Miles thinks the boy is mad, but kindheartedly shows him the deference he demands. Edward falls asleep, and

Miles decides to bring the boy to Hendon Hall, his ancestral home in Kent, where his family will welcome them. He takes Edward's measurements with a piece of string and leaves to get Edward some decent clothes. Edward wakes from a nightmare, thinking he is back in the palace, but realizes that this nightmare is real. A minute later there is a knock on the door; Hugo, a ruffian in his late teens, tells Edward that Miles is hurt and needs him to come to the south end of London Bridge immediately. Edward follows him off."

There are several possible songs in this scene. For instance, Miles's story could be a solo. Edward's nightmare and waking could be a solo as well, although that seems more appropriate for opera than for musical theater. But to us the obvious choice is a comedy duet in which Miles, although he believes Edward is mad, humors him by treating him as the prince and future king he claims to be. There are several good ideas for this song in the novel, and many more could be added.

"*Scene Thirteen.* Inside the Guildhall. Tom is eating a sumptuous meal, sitting at a table surrounded by lords and ladies, with the assistance of many servants. He has not completely unlearned his old habits—he blithely drinks from the fingerbowl—but everyone treats his behavior as normal. A messenger enters and announces that King Henry is dead. All salute Tom as the new king. He immediately pardons the Duke of Norfolk."

Nothing suggested itself immediately in this scene, but there are a number of possibilities for optional songs, depending upon the rest of the score. For instance, there could be a comedy number in which the courtiers obsequiously try to imitate the "prince's" strange eating habits.

"*Scene Fourteen.* Miles's room on London Bridge, and outside. Miles returns to the inn with a set of used clothes and finds Edward missing. As a servant enters with breakfast, Miles questions him and learns that while he was out, Edward was given a message, purportedly from Miles, to meet him at the southern end of the bridge; when Edward left the inn he was followed by a man who, from his description, sounds like John Canty. Miles rushes out of the inn to find Edward, while we hear the news of King Henry's death spreading through the crowd; Edward is the new King of England. End of Act One."

As mentioned before, the end of the first act is a common and obvious place for a song, if it can be justified. Here we envision a short group number in which the people of London repeat the news of King Henry's death—some with sorrow, some with relief—and anticipate the reign of Edward with optimism and support for the new king. This would also make a nicely ironic counterpoint to Miles's search for the missing, mistreated boy. Altogether we have spotted nine or ten places for songs in the first act, and since the average theater score has fourteen or fifteen songs, and the first act is usually the longer one, the number seems about right for this preliminary stage.

Although we did not complete the scenario for Act Two, several song choices did suggest themselves. For instance, there is an essential scene when Canty brings Edward with him to a meeting of his criminal gang. Edward stubbornly asserts his identity, and after some initial merriment the gang leader warns him that they are not traitors; they will not let anyone other than the rightful king call himself King of England, so he must choose another name. One of the gang members suggests "Foo-Foo the First, King of the Mooncalves," and the whole gang takes up the suggestion with delight. This is a good idea for a big group rhythm or comedy number. There would also need to be at least one song in the Hendon Hall sequence, perhaps a solo ballad for Edith, or a duet for Edith and Miles. And as already indicated, there is the moment after Tom denies his own mother, and too late is seized by remorse. We feel that this is the climax of his character arc and one of the emotional high points of the show. We would undoubtedly write a ballad for him at that spot.

THE ORIGINAL

Life on Another Planet, Act One. "Scene One. It is mid-May, in a big chalet-style house overlooking Lake Tahoe, Nevada. Kari Auster, who has just finished ninth grade, is at her fifteenth birthday party. Her mother Terri and her father David are both there, but we soon learn that they are divorced; that David lives outside of Las Vegas, five hours away; and that the house belongs to Terri's boyfriend, Hal. Terri still has her own apartment, but she spends most of her time here, and therefore Kari does too. Both of Hal's children, Carolyn and Mike, are there, as well as Carolyn's husband Brad, friends and business associates of Hal's, and artistic friends of Terri's. There are few children of Kari's age there other than her best friend Jen. For seven years Kari has been living with her mother at Lake Tahoe, and spending holidays, spring breaks and summers with her father in Las Vegas. But according to the divorce settlement between David and Terri, by August 1st Kari must decide whether she will continue to live with her mother at Tahoe, or move to Las Vegas to live with her father for the next three years. She is already feeling pressure about this decision. The usual arrangement is that she spends most of the summer with David, except for the July 4th weekend and the two weeks afterward. But she asks her parents if this summer she can spend two weeks at a time with each of them, to help her make up her mind. Thus she will spend the next two weeks with David, go back to Terri in mid-June, and so on, until August 1st. She feels utterly confused about what to do. A birthday cake is brought in and she tries to blow out all fifteen candles, but she can't. She wonders if that's because she isn't sure what to wish for."

As in *The Prince and the Pauper*, this opening scene seemed a logical place for a number. It might start not at the very top of the show but midway through the party scene, after our characters had been introduced through dialogue. Kari could then give us her perspective on each of them and on her living situation in a solo song. The song would probably continue, interspersed with dialogue and stage business, until the end of the scene, which would tell the audience all it needs to know of her history and the pressure of her upcoming decision. Or this number might be the title song, introducing the central metaphor of extraterrestriality, as Kari describes feeling like an outsider in both her mother's world and her father's. Perhaps it would be even better if the party guests sang certain stanzas as a chorus, and if David and Terri had stanzas of their own, either within the scene, or directly addressing the audience with their own perspectives on Kari and on "the ex." Again, we would probably make this a musical scene built around one song.

"*Scene Two*. The Nevada desert, about halfway between Lake Tahoe and Las Vegas, along the 'Extraterrestrial Highway.' The next day. David and Kari have been driving south, and have just left David's car parked by the side of the road. Kari has gotten her learner's driving permit, and they have stopped for a break, after which Kari will drive for a while. They talk briefly about the summer, but to change the subject Kari starts discussing her interest in Area 51, which is not far away. David notices some unusual plants nearby and leads Kari over to examine them. We see his, and Kari's, love of the desert and of strange and exotic forms of life. David tells Kari how Native Americans used to go out into the desert alone to seek a vision, for insight and wisdom in how to live. He talks about the 'web of life,' the interrelatedness of all things, by which he means life on earth—but which for her includes all life in the universe."

The end of this scene seemed to us a natural place for a ballad, perhaps a solo for David, but most likely a duet which he would start and Kari would later join. With this subject matter, it could make an unusual and effective ballad. It would also tell the audience more about David and Kari, and show the bond between them.

"*Scene Three*. At David's house in Las Vegas, and along the Las Vegas Strip. This is a montage of short scenes over the next two weeks. Kari is staying with David. She is happy to be with him there, or to go with him on field trips into the desert, but when he goes to work she gets lonely. At night she can go into the desert and watch the stars through her telescope, but she spends many days either watching 'infomercials' on television or wandering the streets. She hates Las Vegas and feels strongly that she does not fit in. It's hard for her to imagine living here for the next three years."

We imagined here a sarcastic up-tempo song for Kari about Las Vegas, either a rhythm song or a comedy song. (There have been multitudes of jokes about Vegas, but few theater songs, and even fewer ones with this perspective.) Or this might be where Kari sings a title song, incorporating her feelings about Tahoe and Vegas as well as about her parents and their personalities.

"*Scene Four*. Inside the Flying Saucer Rest Stop, June 1st. This is a diner, souvenir shop and motel in Dry Lake, a small town about halfway along the Extraterrestrial Highway. David and Kari are stopping for lunch on their way back to Tahoe, as they have many times before. It is a colorful place; there is a big sign saying 'Aliens Welcome' at the entrance, and photos of UFOs and other strange phenomena cover the walls, along with eight-by-ten photographs of famous movie monsters. The souvenir shop is at the far end of the dining room, which is less than half full. David offers to help Kari decide where to live by employing logic and a list of pros and cons, but she does not want to discuss it, and soon jumps up to look at the souvenir shop. Zirr, an androgynous individual with strange mannerisms who is browsing the book racks, notices her and observes her closely in a way that might be considered spying. Soon Kari gets bored. She and David leave, as Zirr pays for a pile of books and magazines. Looking out the window after them, Zirr speaks in a strange language into some sort of cellular telephone."

No song ideas jumped out at us in this scene. If Zirr does sing—and we are not sure that Zirr should sing at all—to sing in this scene would definitely be premature. However, there could be a group number about the Flying Saucer Rest Stop, sung by the employees and perhaps the customers as well. It could be either a rhythm song or a comedy song.

"*Scene Five*. Hal's house at Lake Tahoe. This is a montage of short scenes over the next two weeks. Still trying to make up her mind, Kari is afraid that she does not belong here either. Hal is all right but he's not her father, and she can't really talk to him—or to Terri—about her interests; there are always too many people around, including Mike and his jock friends, Hal's suburban bourgeois friends, and Terri's art gallery friends, whom Kari finds pretentious and phony. She likes Tahoe but misses the desert; she likes Jen, but few of the other kids at school. She wonders if she'll ever have a boyfriend. Terri is sympathetic but, as always, the center of a whirl of activity: sculpting, playing tennis, entertaining, networking, spending time with Hal. Kari wonders if her mother really wants her as much as her father does."

We saw a definite need for a song in this scene. It is time for Terri to have a song that helps the audience to understand her, and her relationship to Kari, just as the "web of life" ballad in Scene Two would do for David. We envisioned a charm song for Terri—perhaps a solo, or perhaps with short passages for Kari, Hal, Mike, and Terri's friends. This song would show the whirl of activity that is her natural milieu, in which she remains vivacious and charming while managing her exciting new career success, her new boyfriend, her new home—and, of course, her daughter.

"*Scene Six*. Terri's apartment at Lake Tahoe. June 14th, the day before Kari returns to Las Vegas. Kari and Jen are hanging out and talking, primarily about where Kari is going to live. Then Jen has to go home, and as she leaves, Terri enters with big news: She is going to marry Hal in July. She tells Kari that Hal first suggested the idea after his daughter Carolyn's wedding a few months ago; Terri was reluctant at first, relishing her autonomy after the frustrations of her marriage to David, but she knows that this marriage will be good for her, and hopes that Kari is OK with it. She has essentially been living with Hal for almost a year, and using her own apartment mainly for her sculpting. But Hal is going to convert a shed on his property into an art studio for Terri, so she has decided to give up the apartment. Although this is not a complete surprise, Kari is hit hard. She is not sure how she feels about the marriage, but she hates the thought of losing the apartment, which has been an occasional refuge, a quiet place to study or hang out with Jen. She asks Terri why she has to marry Hal, especially right now—why they can't just keep living together. Even though Kari likes Hal, this would be the final nail in the coffin of her parents' marriage. She doesn't like things as they are, but she's afraid that with the wedding and its consequent changes, her life will become even more unhappy."

We have not indicated it in the scenario, but early in our discussions we had an idea for a ballad for Kari at some point when she is alone in Act One, and this scene might be the right place for it. Our idea was that after Jen leaves, and before Terri enters, Kari has some time alone. (If we did not put a song in this spot, we would definitely have Terri enter while Jen is still there, and Terri's entrance would be part of the reason why Jen leaves.) Kari fantasizes about living among aliens, or being one herself; she wants to believe that on some planets live beings who make smarter choices than people do, and live together more peacefully. We felt that this song is important, not only as a poignant moment, but to show the connection between Kari's unhappy family situation and her interest in UFOs. This is another possible place for a title song. Also, the end of the scene could have a rhythm song, showing Kari's reaction to the news of Terri's wedding. Or the ballad could be moved from after Jen leaves to the very end of the scene.

"*Scene Seven.* Hal's house, the next afternoon. David has come to pick up Kari, who's not home yet. Hal is at work, but Terri tells David about the wedding. He accepts it but still it bothers him, although he doesn't admit it to her. They start talking about Kari's summer and her future, which brings up the past. They start reminiscing, but end up as usual in an argument. Kari comes home in time to hear the end of the argument. David says that there's a lot more to talk about, and insists that when it's time for Kari to return in two weeks, Terri must meet him halfway, in Dry Lake."

We saw a clear cue for a duet in this scene, when David and Terri start reminiscing about their past. It would be a ballad about who they were when they met and fell in love, but also commenting on it from their present perspectives, both bitter and sweet.

"*Scene Eight.* In the car on the way to Las Vegas, later that day. David asks Kari if she has decided whether she wants to live with him or her mother. Still upset by the argument between her parents, she says she doesn't know; what she'd really like is to live with both of them, together. He says that that is never going to happen, and she gets very upset, wildly spinning a fantastic picture of the three of them living together in harmony with nature and each other, halfway between Tahoe and Vegas, finding all they need in the desert, where they can sculpt, research, and look for alien spaceships."

We saw Kari's outburst as a cue for a musical scene that would be a sort of "Mad Song" for her. We felt it important to show that she feels her already unsatisfactory world is coming apart, which becomes part of her motivation for running away. Her fantasy would be primarily about the three of them, and might or might not mention UFOs.

"*Scene Nine.* David's house in Vegas. Early morning on July 4th, the day Kari is to return to Tahoe. Kari comes out of her room as David is leaving for his daily run. When he is gone, she brings out a big new hiking backpack, which she starts to pack with outdoor gear and food. She calls Jen at Lake Tahoe and says her parents are still fighting about her, and she can't fit in at either place. She tells Jen that she is running away; she doesn't know where she'll end up, but right now she's going into the desert to seek insight and wisdom. She needs time and solitude to figure out where she belongs. She can't live with either parent; neither one really understands her, and she doesn't feel at home in either Tahoe or Vegas. She does not pack her telescope."

Here again we had a clear idea for a song for Kari: an angry goodbye to her parents, expressing her resentment and her need to "get away." It would be a driving, up-tempo rhythm song.

"*Scene Ten.* A street in the center of Dry Lake, later that day. David and Kari are sitting on a bench, waiting for Terri, who is late as usual. Kari tries to convince David that it's all right for him to leave without waiting for Terri to show up, but he refuses to leave her there alone. She tries to pick a fight with him and, when that doesn't work, sends him down the street to get her something to eat. As soon as he is gone, she hoists her backpack and heads out into the desert. End of Act One."

We did not have any song ideas for this scene. But we didn't think it necessary to have a number here, since Kari has already made clear in the previous scene what is about to happen. Some ominous music at the end could be enough, as it is at the end of the first act of many musicals such as *Guys and Dolls* and *Camelot.* However, another possibility would be to save Kari's rhythm song from the previous scene for this scene. We could establish as a convention that while Kari is belting out her big number, the passersby don't hear her; but our inclination would be to set the song a little more realistically. In any case, we have spotted eight or nine places for songs in the first act, which is a reasonable number. Since they might not all turn out to be good choices, we would probably try to add another one or two if this were to be a complete show.

Although we did not complete the scenario for Act Two, several song choices already suggest themselves. For instance, we feel that our first scene with Kari alone in the desert, confident that she has packed everything she needs to survive on her own, is a perfect spot for a comedy song, which we will discuss shortly. Zirr might or might not sing; but certainly Kari could have a song showing her awed reaction to Zirr's pithy words, or her growing certainty that Zirr is an extraterrestrial, or both. Obviously there would also have to be at least one or two numbers for David and Terri, alone or together. We have not thought through the plot of Act Two in enough detail to know more, but there are likely to be additional complications relating to Terri's relationship with Hal, or scenes with Hal's children and son-in-law, that might provide additional perspective on families and relationships. We also feel that the final bittersweet resolution is another place for a song in which Kari accepts that her parents will not get back together and that she'll be okay, although without further development we don't know whether it would be a new song; it could be a reprise of David's ballad from Act One, Scene Two about the "web of life" and the interrelatedness of all things.

THE NEXT STEP

Once the score has been spotted, the songwriters need only choose a song and start writing. Librettists usually prefer to write the book in order, but there is no necessity for the score to be written that way.

According to Fred Ebb, when he and John Kander wrote a score they usually started with the opening number, because they needed to define the show before writing the rest of it. But most Broadway veterans prefer to write the opening number, if there is one, last. As discussed in chapter 3, the opening is usually the most crucial number in the show because it sets the scene, the tone, and the style of everything to follow. The story of the first production of *A Funny Thing Happened on the Way to the Forum* makes clear that no matter how good the rest of the show may be, it will not work if the opening number gives the audience a false idea of the show it is about to see.

It's often very difficult to give a proper setup for the show, however. And this is especially true when the rest of the show has not yet been written. Many things may change in the writing, and a whole show often has a different effect from the sum of its parts. This might explain why opening numbers tend to be rewritten or replaced more than any other songs in a score. The longer you live with your characters and your material, the better you'll understand them and understand what the show needs to introduce it properly.

Some songs will almost inevitably have to be changed or replaced during the writing or the production of the show, especially those we have called optional songs. Early in the writing of *My Fair Lady*, for example, Alan Jay Lerner and Frederick Loewe played their first five songs for Mary Martin in the hope that she might play Eliza. Of those five songs, two were eventually cut and another ended up later in *Gigi*. (Mary Martin's reaction to the other two songs explains why she never played Eliza: She felt that "Just You Wait" was stolen from *Kiss Me, Kate*'s "I Hate Men," and that "The Ascot Gavotte" was "just not funny at all.") A third of the songs that Stephen Sondheim wrote for *Forum* were cut, and a similar percentage of Bock and Harnick's songs for *Fiddler on the Roof* were cut as well. Since this is a fact of theatrical life, it should not discourage you from proceeding with your favorite ideas for songs. If you correctly grasp the emotional core of your story from the beginning and musicalize its high points, at least some of those songs will remain.

For this book, we decided that we would write one song of each of the four main types discussed in chapter 5: ballads, comedy songs, charm songs, and rhythm songs. We wanted to give equal time to each of our two projects, so our first step was to choose four songs for each of the projects, from which we would then pick two. Of course, in writing a real show you would never work this way. Nevertheless, it may be helpful to discuss some songs we didn't write, as well as those we did.

For *The Prince and the Pauper*, we decided that our ballad would be Alice Canty's song from Act One, Scene Ten. As we mentioned, this ballad would show her love and concern for Edward, whom she believes to be her son Tom gone mad. This is one of the four songs we went on to write, so we will discuss it further below.

For comedy we picked the song we spotted in Act One, Scene Twelve, when Miles treats Edward as the prince he claims to be. The good-hearted Miles believes Edward to be a poor lunatic and feels pity for him as well as admiration for his bravery and wit. Having decided to take the boy home and care for him, Miles shows him all the deference customary to the Prince of Wales, even to the extent of not eating or sitting down until the boy is served. Miles is also amused by the absurdity of the situation, as he sees it: he, a member of the lesser nobility, bowing to a poor crazy beggar boy. We think this song could be both charming and funny, and the title would be "Yes, Your Majesty." (In our adaptation, Edward does not yet know that he is king—but he expects that he will be soon. So before the song starts, he would suggest that Miles address him as befits a king rather than a prince.) The song would end with Edward granting Miles permission to sit in his presence at any time, as he does in the novel. It would also allow us to use a short reprise of this song to conclude the show with a laugh. As in the novel, Miles would sit down in front of the new king, and the horrified nobles would want to run him through with their swords. But Edward would explain that Miles has his permission to sit, which Miles would do again with a roguish smile and a final "Yes, Your Majesty."

For a charm song we decided on Act One, Scene Three, where the prince and the pauper trade stories of their daily activities, and each envies the other's life. This is an extremely charming moment in the novel. In a musical the audience must care for the protagonists, and since the function of a charm song is precisely to show a character as appealing and likeable, this seemed a perfect place for one. Since it is also the situation that triggers the rest of the story, it is a natural moment to be musicalized. This is one of the four songs we went on to write, so we will discuss it further below.

For a rhythm song we decided on Act One, Scene Five, when Tom first finds himself alone in the prince's chambers. As we suggested earlier, in the first part of the song Tom might start by fantasizing excitedly about what the prince might bring him: wonderful food . . . or the gift of a lovely suit of clothes . . . or perhaps a position in the palace, such as "Companion to the Prince of Wales"? Tom's imaginings could become more grandiose until, in an interlude, he would realize how long the prince has been gone, and open the door to look for him. Seeing the armed guards and misunderstanding why they are there, in the last section of the song he might start imagining that they are bringing not gifts and rewards, but punishments and beatings.

For *Life on Another Planet*, we decided that our ballad would be the duet for David and Terri that we spotted in Act One, Scene Seven. As we indicated, this would be a ballad about the way things were when they met and fell in love, which would also fill in their "backstory" for the audience. But it would not simply be a sentimental journey. While remembering how idyllic things seemed at the time, they would also comment on it from their present perspectives. Thus it would be a "Sorry-Grateful" type of song: nostalgic in some places, but also dryly sarcastic and rather sad in others.

For a comedy song, we chose an idea we had for Act Two. We have not shown any details of our Act Two scenario, but this would occur in Scene Two, which would show Kari alone in the desert, a few hours after having left Dry Lake. We thought it would be funny to see this fifteen-year-old begin her wilderness trek with great confidence that she can survive in the desert on her own, only to find that she may have underestimated what was involved. This is one of the four songs we went on to write, so we will discuss it further below.

For a charm song we decided on Act One, Scene Five, which shows the whirl-wind of activity surrounding Terri at Lake Tahoe. If Scene Two is the scene in which we really get to know David, then this is the scene in which we really get to know Terri. At this time in her life she is so busy that she seems less interested in Kari than David does. We felt a charm song would help the audience under-stand and like her, despite her apparent obliviousness. People like Terri who are creative, sunny, scattered, and always occupied can be charismatic and fascinat-ing, but to be the teenage child of such a person can be tough. Showing Terri in the midst of her tempest of activities, and trying to balance them all—working, playing, spending time with her boyfriend, networking, entertaining—would help to explain why she seems to have less time for Kari than David does.

For a rhythm song, we chose the moment from Act One, Scene Nine, when Kari expresses her desire to get away from both her parents and the impossible choice she has to make. This would be the emotional climax of Act One, and seemed clearly to demand an angry up-tempo rhythm song. Since it is one of the four songs we went on to write, we will discuss it further below.

From these eight songs, we chose four—one of each functional type—and to keep things balanced, we selected two from each of the two projects. We decided on the ballad and the charm song from *The Prince and the Pauper*, and the comedy song and the rhythm song from *Life on Another Planet*. We have completed what we consider to be the two most difficult types of songs, the bal-lad and the comedy song, and begun the other two. The point is not to show off the songs we wrote, of course, but to help you write your own.

Writing a Ballad

As discussed in chapter 5, the essence of a ballad is the expression of strong emotion. The first step, then, in writing any ballad is to decide the character (or characters) who will sing it, and the second step is to decide, as precisely as pos-sible, the emotion you wish to express.

After that comes the third and hardest step—finding the language in which to express the emotion. Simply to express an emotion baldly makes a lyric that is both clichéd and uninteresting. Imagine a love song whose lyrics consisted of nothing but:

I LOVE YOU.
I LOVE YOU SO MUCH.
YOU MEAN EVERYTHING IN THE WORLD TO ME.
I LOVE YOU MORE THAN I'VE EVER LOVED ANYONE ELSE.
I CAN'T LIVE WITHOUT YOU. (etc.)

One of the wonderful things about drama, or for that matter about being human, is that people use so many different ways to express similar emotions. These differences can stem from such factors as upbringing, social mores or constraints, and personal habits or inhibitions. One crucial factor is the history of the relationship between the person(s) talking or singing and the person(s) addressed; another is their situation at that particular moment. It is from these factors that the lyricist can find the materials for a lyric that is both expressive and interesting.

In the situation we have chosen from *The Prince and the Pauper* we have some very specific guidelines, because Act One, Scene Ten of our adaptation takes many details from chapter X of Mark Twain's novel. In this chapter Canty brings Prince Edward to Offal Court. When Edward asserts his identity, Mrs. Canty is horrified, thinking that her son has gone mad. She exclaims, "O poor Tom, poor lad!" and "O my poor boy!" as she tries to jog his memory. When he tells her that he has never seen her before, she bursts into tears. After Canty orders everyone to go to bed, she goes back to the boy, strokes his hair, offers him food, and hugs him repeatedly before going back to her own corner.

For the bulk of our ballad we chose to musicalize Alice's feelings of concern, love, and helplessness, concentrating them in a soliloquy sung after she has gone back to her corner. We chose the title from her exclamations, which suggested the situation without being specific too soon. We considered "My Poor Boy," or "My Poor Little Boy," but we decided not to use the word "my" for reasons that we will explain below. We might have used "Poor Little Boy," but instead we chose "Poor Little Lad," because "lad," a word that has fallen out of general use, suggests a bygone period. While there are many other possible titles for this situation, "Poor Little Lad" provides a common jumping-off point for each stanza.

In writing any lyric, it is valuable, if not essential, to start brainstorming or free-associating at an early stage, collecting and writing down lyrical ideas. These ideas may include possible titles, appropriate words and phrases to express the character's emotion, rhymes that seem suitable for the character and the subject, and any ideas for structure that come to mind. Stephen Sondheim has said that he also tries to write a statement in prose of what the song is about, which is especially useful if you do not have a collaborator with whom to discuss the song.

For this song, we started writing down thoughts that might be running through Alice's mind. Here are some of the thoughts that occurred to us:

What's happened to you?
Where is your poor mind wandering?
It breaks my heart to see you like this.
You've been hurt so badly already.
I wish I could help you.

I don't know what to do.
I know how to take care of cuts and bruises, but not this.
How can I banish what isn't really there?
What can I do to make the bogeymen and phantoms vanish?
You're lost in some crazy dream.
But I know your pain is real.
You no longer know who I am.
You must be my son. You have to be!

As you continue to collect lyrical ideas, a structure should start to become clear, a way to organize the ideas both psychologically (as a believable sequence of thoughts and emotions) and formally (as a lyric). In this case, we conceived a first stanza expressing Alice's heartache at the boy's condition, and a second stanza expressing her wish to help. This suggested two A sections, which in turn suggested an AABA structure. We could see a B stanza about how she has always been able to heal his cuts and bruises, but has no idea what to do about his madness. The fourth stanza would be another A section. At this point we wanted to suggest that she is already starting to have doubts that the boy is her son. These doubts would prompt the insistence that he is, that he *has* to be. (In the novel Mrs. Canty says "He *must* be my boy!" even after performing the test that proves he is not. We liked the thought but moved it earlier, where we felt it was more appropriate.)

By now we had another structural idea, because Twain's focus on Mrs. Canty continues beyond this point. The feeling that there is something different about Edward, the nagging thought that he might not be her son, torments her until she devises a test. From a traumatic event in infancy, Tom's consistent response to any sudden fright has been a reflexive pushing-away gesture. She tests the boy with a sudden noise and a light; he is startled, but he does not react as she knows Tom would. (She performs this test three times, which we think would be redundant on stage; one test is enough.) She goes back to her corner and falls asleep, even more distressed. Since this later moment clearly continues Alice's earlier inner monologue, we wanted to integrate it into the song.

We decided that Alice would get up and go over to Edward after the fourth stanza, while the music continues. We didn't want to have her sing during the test, so we decided that it would be accompanied by instrumental underscoring, after which she would sing a final A section. In this stanza, Alice would now have serious doubts that the little boy in her room is Tom—but she would still feel concern for him, whoever he might be. (This is why we decided not to use the word "my" in the title; it would not work in this last stanza.) However, it was also clear to us that as a loving mother, her last and most poignant thought would be not about him, but about her actual son—because if this boy is not Tom, then Tom never came home, and he must be in desperate straits somewhere else.

Now that we had a clear idea of the overall structure, and of what each stanza would be about, we could start the task of finding a meter for each line and filling in the actual words. Here is our scene and the lyric for "Poor Little Lad."

SCENE TEN

(The Canty rooms. All are asleep except Alice, a candle burning by her head.)

ALICE
(She is tossing and turning, muttering to herself. Then she suddenly sits up.)
There is something about him! Something—different . . . Something in his eyes . . . in his voice . . . But he is afflicted, poor lad . . . But surely even madness could not change the lilt of his voice, the look of his glance . . . Could it be, could it possibly be?
(vehemently)
No, no, he is my own dear Tom. Of—of course he is.
(sings, glancing occasionally at Edward)

POOR LITTLE LAD,
SO LONELY, SO LOST AND CONFUSED.
TO SEE YOU THIS WAY TEARS AT MY HEART.

POOR LITTLE LAD,
I KNOW YOU'VE BEEN HURT AND ABUSED.
I WISH I COULD HELP YOU, BUT WHERE DO I START?

I KNOW HOW TO CARE FOR CUTS AND SCRAPES,
BUT I CAN'T BANISH PHANTOMS WITH SHIFTING SHAPES.
THE PAIN YOU SUFFER IS REAL,
BUT THIS IS A HURT I CAN'T HEAL.

POOR LITTLE LAD,
YOU DON'T SEEM TO KNOW ME AT ALL.
BUT KNOW ME OR NOT,
YOU'VE GOT TO BELONG TO ME.
IF YOU WEREN'T MINE,
THEN WHOSE LITTLE LAD WOULD YOU BE?
(Music continues under.)

If you weren't mine . . . But how can I know? How can I be sure?
(a beat)

The firecracker! It blew up in his face when he was barely two. And since then, whenever he has a sudden shock . . .

(She picks up her candle and goes over to the sleeping
Edward. Quietly)

If you are my Tom, you'll throw out both your arms to protect your face, as he always does. And if you don't . . .

(She thrusts the candle at his eyes and raps the floor loudly.
His eyes fly open and he starts to rise. Quickly)

No, no, my lad, 'tis nothing. 'Tis nothing. Go back to sleep. Sleep.

(He subsides and falls asleep. She returns to her bed, even more distressed.)

POOR LITTLE LAD,
SO LONELY, SO LOST AND CONFUSED.
YOU COULD BE MY SON,
BUT ONE THOUGHT IS HAUNTING ME:
IF YOU AREN'T MINE—
OH GOD, WHERE CAN HE BE?

While much of this lyric is taken from our collection of ideas, a few of the changes are worth noting. For instance, the line "It breaks my heart to see you like this" has undergone two changes to become "To see you this way tears at my heart." First, the word order has been reversed; it seemed to us that a mother would be as likely to start with the thought of the boy's condition as with the thought of her own pain. Second, we have changed "like this" to "this way," a small change but a definite improvement. Because of the short vowel and the final sibilant, "this" is not a good sound for a long note; it does not sing well, especially in a ballad. "Way," an open diphthong with no final consonant, is better.

If we were going to continue working on this score, we would probably consider the further revision of some of these lines, such as "But I can't banish phantoms with shifting shapes" in the third stanza. It was difficult to find words with which an ignorant sixteenth-century woman would describe madness, words that would be both lyrical and believable. We like her idea that in his delusion the boy must see "phantoms" and things that are hazy and fluid rather than clear; but we worry that "shifting shapes," which was chosen at least in part for a rhyme, is a little too much. The alliteration might suggest education and self-consciousness, which are inappropriate here. In addition, there are a lot of sibilants in this line, including three uses of "sh." On the other hand, the alliteration also gives a rhythmic impetus to the line, which would be missing if we used something like "hazy shapes."

We felt more confident about the internal rhymes in the middle of the fourth and the last stanzas, *not/got* and *son/one*. In these places we felt that the disadvantage of internal rhymes—that they usually connote wit and erudition—is outweighed by the rhythmic thrust they impart, especially because they are simple one-syllable rhymes that do not call attention to themselves.

We would also direct your attention to the final line of the first chorus. It's always effective to have the end of a lyric bring back the title, or a variant of it, in some emotionally appropriate way. To repeat "poor little lad" literally would not have worked, but we liked being able to use "whose little lad" at the end of the first chorus. We were strongly tempted to do the same thing at the end of the song, and make the final line "Then where can my little lad be?" But here we felt that it was important for Alice to end with a direct, naked utterance, and to use "my little lad" rather than the simple "he" would seem artificial and inappropriate for the character. Fewer syllables in the line also seemed more apt for the feeling of a final heartfelt cry, as did the addition of "Oh, God."

In our preliminary discussions of the music, one important question was our conception of Alice's vocal range. As mentioned in chapter 6, composers should be flexible about keys when first writing a song. This does not mean, however, that they cannot decide on a particular type of voice for a character, if it seems integral to the role. Some women's roles, for instance, might demand a belt voice, while others might call for a legitimate soprano voice. This can always be changed, of course, but the clearer the composer's choice is from the beginning, the fewer and smaller any adjustments will need to be. Suppose we started with the idea that Alice was a belter with a strong low tessitura, and later changed our minds and made her a soprano. If we had already written the song with important syllables set to low notes, we could transpose it to a higher key for a soprano—but it would still be less effective, because sopranos' low notes are usually weak. In this case, we decided that Alice should be a mezzo-soprano with a "mix"; neither a belt voice nor a soprano voice seemed appropriate to us. We have put the song in a key that seemed right for that type of voice. Of course, even if the role were eventually cast with the type of singer we had in mind, we might still need to move the key up or down a little for the singer. And if we had to sing it for a producer we would also transpose it for that occasion, to fit our own voices better.

Another consideration is the overall range of the song. As mentioned in chapter 6, a comfortable average range for singers is about an octave and a half, although professionals should have a range of at least an octave and a sixth. If your song's range is much wider than this, you had better have one or more specific performers in mind—and know that at least one of them will be available and willing to play the role when the time comes. But veteran songwriters rarely need to worry about this, because a sense of practical vocal range becomes ingrained in their subconscious with experience. Also, most good melodies tend to move by steps rather than leaps, so it would take an unusually wide-leaping melody to demand a range wider than an octave and a sixth. For both of these reasons, experienced songwriters often do not worry about range until after the song is written, at which point they can make small adjustments. As you will see, the melody turned out to have a range of only a ninth, so there was no need for any adjustments.

Another aspect of the music to ponder in advance is the style. If a composer is so limited as to have only a single style this may not be a question, but for composers with a greater range, choices need to be made, in regard to both the score as a whole and the particular song. We had already decided that the score as whole would be essentially contemporary but with a slight period flavor. For this ballad we decided that there would be no strong rhythmic pulse, and that we would try to use or to suggest a mode such as the Dorian. (The Dorian mode is the white-note scale beginning on D, which differs from the minor scale in having a whole step rather than a half step between the fifth and sixth notes. It can be found in old English folktunes such as "Greensleeves" and "Scarborough Fair.") However, there was no need to stay with a single mode, and ultimately we also made use of the Aeolian mode (equivalent to the natural minor scale) and modal mixture.

Even before we completed the lyric we had in mind a beginning like one of these, in A minor or A Dorian.

By the time we finished the lyric, however, it occurred to us to try a different rhythm—making "poor" longer—and a different direction for the melody, such as one of these:

We eventually decided against both of these as well. We liked a slight pitch rise on "little," which made for a motif more distinctive than the original one; but we felt that in both of the above versions the rise was a little too much, and both put the word "little" (which we consider the least important of the three) into too strong a relief. Finally we arrived at a version of the motif that satisfied us, and we completed the song.

Poor Little Lad

Words and Music by
Allen Cohen and Steven L. Rosenhaus

Poor lit-tle lad, So lone-ly, so lost and con-fused. To

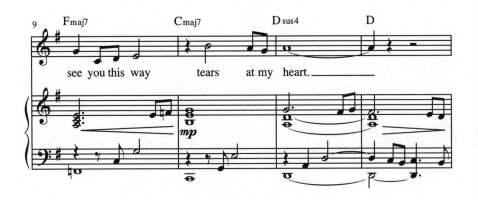

see you this way tears at my heart.

Poor lit-tle lad, I know you've been hurt and a-

bused. I wish I could help you, but where do I

start? I know how to care for cuts or scrapes, But I

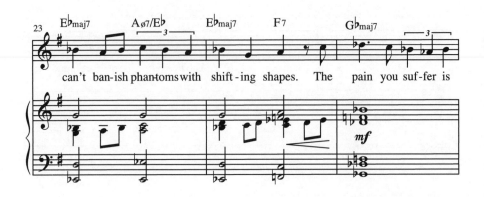

23 E♭maj7   A∅7/E♭   E♭maj7   F7   G♭maj7

can't ban-ish phan-toms with shift-ing shapes. The pain you suf-fer is

26 Fm2   Fm/E♭   Dm9   Em7

real, But this is a hurt I can't heal.

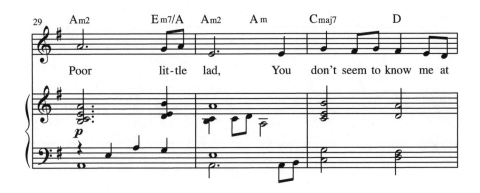

29 Am2   Em7/A   Am2   Am   Cmaj7   D

Poor lit-tle lad, You don't seem to know me at

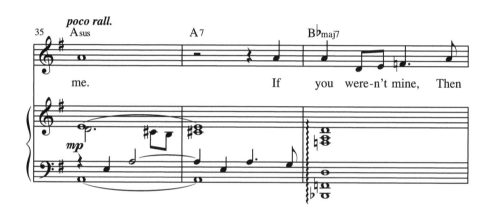

END:

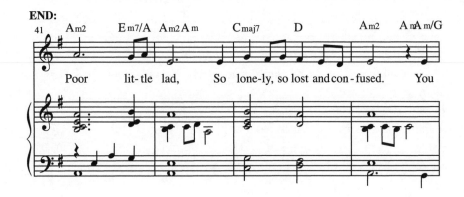

Poor lit-tle lad, So lone-ly, so lost and con-fused. You

could be my son, But one thought is haunt-ing me: If

you are-n't mine– Oh God, where can he be?

We would like to discuss a few points about this setting. First, the choice of material for the four introductory bars is almost purely arbitrary. The actual music of the introduction would be dictated by aspects of an actual production: the direction, the actor's choices, and the given instrumental forces.

In our mixing of modes, we occasionally created cross-relations within stanzas, as for instance following F-sharp of measure 15 with F-natural in the piano part of measure 16. In some scores this would sound strange, but here we felt it gave an appropriate period flavor. The same reason underlay our total avoidance of the leading tone (the raised seventh degree of the scale, which in this song would be G sharp). Every time the melody resolves to the tonic A, as for instance at the very end, the previous note is always a G natural rather than a G sharp, and the harmony is an E minor seventh chord rather than an E dominant seventh. This modal sound helps to maintain the period flavor.

The music evolved as we worked on the song. For example, the original melody for the words "got to belong to me" in measures 34 and 35 (and for "one thought is haunting me" in measures 46 and 47) was:

This is stronger than the final version because it goes to a higher note, but we decided that the final version was better, because the real climax of the stanza was "whose" in the last line. To go up to the C on "belong" would anticipate the C on "whose" and thus diminish the effectiveness of the last line.

We also want to point out two apparent errors in setting the words, which appear in measures 37 and 49. Most contemporary Americans pronounce "weren't" and "aren't" as two-syllable words. Contemporary English people pronounce them as one-syllable words. We decided to use two syllables here because we are Americans, writing for American performers and American audiences, and because we don't know how English people pronounced these words in 1547. If this show were being produced in Britain, we would keep the two-note settings, but make them melismas on one-syllable words.

Writing a Comedy Song

As discussed in chapter 5, the function of a comedy song is to make the audience laugh out loud. The starting point for a comedy song, then, is finding ideas, situations, or things to say that are funny enough to provoke laughter. Different people find different things funny, but bear in mind that theater is a popular art form, and for a comedy song to work, it needs to seem funny to most of the audience. If you are writing a song for a show that will be performed for a particular audience, you can write jokes that only they might understand, but otherwise it is best to avoid in-jokes. Similarly, many people find puns not only unfunny but repellent. To count on puns for laughs is usually a mistake.

We thought a good spot for a comedy song in *Life on Another Planet* was the second scene of Act Two, which shows Kari alone in the desert for the first time. We thought of items that she might pack in order to run away: a compass, a flashlight, water bottles, energy candy bars, trail mix, long underwear, a first aid kit, binoculars, tampons, toilet paper, toothbrush, soap, and so on. We tossed around ideas for laughs that could come from the things she might have packed or forgotten to pack: There might be humor in her remembering to pack something like a toothbrush but forgetting to pack toothpaste, or bringing a flashlight but forgetting to bring batteries. But many of these ideas seemed improbable for a girl who, although young, was highly intelligent and had been in the desert many times. If anything, Kari would be likely to have made a list of what she would need, and to have checked it before leaving David's house. Then we hit upon a variant of this idea that would allow Kari to have made a list. It necessitated a subsidiary plot element that would be introduced early in the show, in order to have its payoff in this scene.

Act One, Scene Three is the first time Kari is shown in David's house, where she is often lonely and bored. We decided that while she is watching infomercials on television, there could be one particular commercial for a sort of super Swiss Army knife combining a vast array of blades and other utensils. The announcer's voice-over could be so excited and exaggerated in its claims for the tool as to be funny in itself. In Scene Nine we would see that Kari has actually bought this tool, which is about as big as two fists together. It would stand at the top of her list, so she would pack it before anything else. She couldn't pack all the gear she would need for her trek because David would notice, so this tool would be a great space saver. Then during her phone call to Jen, while discussing her preparations for the trek, she would get excited about how "cool" the new gadget is, and pull it out of the backpack to better describe it to her friend.

As discussed in chapter 5, there are essentially two types of comedy songs: those that build to a single long joke, and those with a series of smaller jokes. Long-joke songs are much rarer. The problem with the long-joke song is that if the one joke does not work at the end, then the whole song does not work. Multiple-joke songs have a different, if less serious, problem: If the jokes do not get bigger and funnier, then the audience often feels disappointed at the end.

We decided that in this song each stanza would pose a different problem for Kari. Originally we hoped that the solution at the end of each stanza would bring up a new problem in the next stanza—for instance, she needs a knife to cut something, but then realizes she needs a screwdriver to open the box containing the knife, and so on—and perhaps the solution for the last stanza would turn out to be the item missing in the first stanza, leaving a circle of unsolved problems. But we found no way to make that believable, nor could we think of subsidiary jokes that would help build toward the final one. Also, if each stanza had a different "solution," there would be no way to link each one to a title or central concept. Our first thoughts for a title were of the "I can handle this" variety, such as "No Problem" or "I've Got It Covered." But as the idea of a bunch of different tools coalesced into the notion of a single, versatile tool, we decided that our "Swiss army knife on steroids," which was sold by an outdoor gear company that we named The Great Northern, would be called the "Great Northern All-in-One." With this as our central idea, it seemed clear that all the problems would require the All-in-One as their solution. This seemed to demand a long-joke structure.

Kari would be confident that she has been careful and thorough in her preparations, and so at first she would be unfazed by problems. Clearly these problems should be immediate survival issues such as food, warmth, and shelter. Thus we arrived at a structure in which the first A section expresses Kari's need for a fire, the second her hunger, and the third her need for shelter. At the end of each stanza would come her confident assertion that the Great Northern All-in-One can solve her problem. This suggested that each A section would itself consist of a smaller verse-refrain form, with the verse expressing the problem, and the refrain extolling the virtues of the tool. Somewhere in the song she would also express her excitement about her adventure and her confidence that she can do fine in the desert alone, to better set up the ending; a release section seemed a logical place for this. We also decided that we needed to add a tag at the end, and a short introductory verse before the problems start.

One final word before we present the scene and the lyric for "The Great Northern All-in-One": While there are no period considerations in a contemporary setting, there is a different problem in the use of current slang. Since today's slang may well be outdated tomorrow, we decided to keep the slang to a minimum of commonly understood expressions.

SCENE TWO

(The desert, shortly before sundown. A rocky landscape with small hillocks on either side and a sandstone shelf at the back, creating a somewhat protected area. Scattered bushes and brush around.)

(Kari enters, carrying her enormous backpack, tired and hot. She slips the backpack off her shoulders and lets it slide to the ground. She totters for a few seconds . . . then her feet give way and she slides down to the ground, leaning against the backpack.)

KARI

Yeah, let's stop here.
(She unclips her water bottle from the backpack and takes a long drink, then looks around.)
This seems to be far enough for the first night. Got here just in time, too—I can set up camp before it gets dark.
(She stands up slowly and stretches.)
Jeez, that pack is heavy. Next time, I'll get something with wheels . . . like a shopping cart.
(thinking about what she just said)
"Next time"? Uh, no, I don't think so. This is it. A few days or a week out here, figure out where to go next, and then go. So long, farewell. No problem. I've got this all planned, down to the last detail.
(opening up the backpack, she sings)

I MADE A LIST, I CAME PREPARED;
I DID IT RIGHT, AND I'M NOT SCARED.
EVEN THOUGH I'M OUT HERE ALL ALONE,
I'VE GOT EVERYTHING I NEED.

JEEZ, IT'S GETTING COLDER . . .
THAT WIND HAS REALLY STARTED BLOWING.
BUT I SEE PLENTY OF BRUSH AROUND,
SO IT'S TIME TO GET A FIRE GOING!

(taking out a small hard case and trying to open it)

THERE'S LOTS OF MATCHES IN THIS CHILDPROOF BOX—
AND IT'S JAMMED TIGHT SHUT! WHAT A DRAG.
BUT I'VE GOT THE ANSWER TO THE PROBLEM
RIGHT HERE IN MY BAG!

(rummaging in the backpack)

IT'S THE GREAT NORTHERN ALL-IN-ONE TOOL,
SO VERSATILE AND OH, SO COOL!
IT CAN GUT A FISH AND CLEAN THE SCALES,
IT CAN FIX A FLAT, AND THEN TRIM YOUR NAILS!
THE BEST TOOL UNDER THE SUN
IS THE GREAT NORTHERN ALL-IN-ONE!

Now where is it? Oh, come *on* . . . Yeah!
(She reaches in deep and pulls out, not the All-in-One, but a bag of trail mix.)
That's not it—but, come to think of it . . .

BOY, I'M GETTING HUNGRY . . .
SO LET'S BEGIN THE DINNER PARTY.
THERE'S POUNDS OF TRAIL MIX IN PLASTIC BAGS,
BUT I FEEL LIKE SOMETHIN' HOT AND HEARTY!

(She pulls out a can of food.)

AH, PORK AND BEANS, WE'RE TALKIN' CAMPFIRE FOOD—
BUT THE POPTOP'S MISSING ITS RING!
THAT'S FINE, THOUGH, 'CAUSE HERE INSIDE MY BACKPACK
I'VE GOT JUST THE THING.

(rummaging in the backpack again)

IT'S THE GREAT NORTHERN ALL-IN-ONE TOOL,
SO VERSATILE AND OH, SO COOL!
IT CAN OPEN ANY CAN WITH EASE,
AND IT'S EVEN GREATER AT GRATING CHEESE!
THE TOOL THAT'S SECOND TO NONE
IS THE GREAT NORTHERN ALL-IN-ONE!

(stopping her search to look around)

WHAT A RUSH TO BE OUT HERE
WHERE I CAN THINK AND BREATHE FREE!
I CAN DO THIS, I DON'T NEED ANYONE—
ANYONE BUT ME!

(She suddenly yawns.)
Wow, I'm really tired. Maybe I'll just eat in the morning.
(She takes out a sleeping bag and plastic groundcloth.
Distant thunder and lightning.)

HEY, DID I HEAR THUNDER?
IT LOOKS LIKE THERE COULD BE A RAINSTORM.
IT SUCKS TO SLEEP IN THE RAIN ALL NIGHT—
BUT HOLD ON, I THINK I'VE GOT A BRAINSTORM!
TO MAKE A SHELTER FROM THIS PLASTIC TARP
YOU WOULD NEED SOME SERIOUS GEAR.
BUT I'M GOOD, I'VE GOT A SECRET WEAPON
AND IT'S RIGHT IN HERE!

(She rummages around fruitlessly.)

SOMEWHERE RIGHT IN HERE . . .

(emptying everything out of the pack onto the ground)

WITH THE GREAT NORTHERN ALL-IN-ONE TOOL,
SO VERSATILE AND OH, SO COOL,
I CAN CUT SOME BRANCHES, DIG FOUR HOLES,
EVEN NOTCH THE BRANCHES TO WORK AS POLES!
AND NOW TO GET THE JOB DONE . . .

(But now the pack is empty, its contents strewn over the ground.
Thunder and lightning again, now much closer.)

OHMIGOD, WHAT A BRAIN-DEAD FOOL!
I CHECKED MY WHOLE LIST,
BUT THE ONE THING I MISSED
WAS THE GREAT NORTHERN ALL-IN-ONE TOOL!

(Blackout.)

Since this is a long-joke song, everything depends on the audience's accepting that Kari, despite her list, forgot to pack the All-in-One tool. We felt that in Act One, Scene Nine, it would be plausible for Kari to bring the tool out of her backpack to describe it to Jen over the phone, and that the conversation might suddenly turn toward her feelings of desperation, so that she might carelessly put the tool down on a chair. While most of the audience might not notice this, at the end of "The Great Northern All-in-One," some might remember that Kari had taken the tool out and never put it back. Thus it would make sense that she had forgotten the tool and left it behind.

The crucial elements of a comedy song are the idea and the lyrics. In this case, the lyrics have little humor in and of themselves; the success of the song depends on whether we have sufficiently set up Kari's character and situation so that the idea of the ending, and the stage picture—the forlorn Kari, her belongings strewn about her, a downpour about to start right overhead—are funny. We did try for a little wit in the first two refrains, and in the "And it's right in here!/Somewhere right in here . . ." of the last verse, but that's about all. In a one-joke song, the bridge only needs to be serviceable; if the song works it's fine, and if not, it doesn't matter. If we went on to revise the song, we might attempt to "punch up" some of these sections, especially the bridge, with a few extra laughs.

In our preliminary discussions of the music, we decided that Kari's voice might be lower and more "belty" than Alice Canty's. While there are eighteen-year-old singers with trained legitimate voices, in our experience more young singers have rock or belt voices, and such voices tend to be more powerful than the legitimate ones. We also felt that a belt voice was more appropriate for a contemporary fifteen-year-old than a legitimate or high voice. The upper range of this song is close to that of "Poor Little Lad," but it goes lower, and the overall sound is quite different.

We knew that the score as a whole would need a reasonably contemporary style. For this particular song, we decided on a light rhythmic feel for the verses, almost like a charm song, and a slightly stronger accompaniment for the refrains. We chose a swing or shuffle rhythm, partly to get the charm-song feel, and partly as a contrast with the other songs we were writing. (Swing rhythm is actually a compound meter in which a four-beat measure is played as 12/8 rather than 4/4; in other words, each beat is divided into three equal parts rather than two. But by convention, it is usually written as 4/4, with unequal divisions of the beat written as dotted eighths and sixteenths.) We also decided that the verses should start in minor, expressing the problem at hand, and the refrains should be in major, expressing Kari's confidence in the All-in-One.

We structured the melody so that each verse would start low and rise, reflecting Kari's mounting excitement. The refrains would start high with the line "Great Northern All-in-One tool," dip lower in the middle, and then end high with the triumphant last line. As for the tag, after a great deal of discussion we decided that it should start high but end low.

The Great Northern All-in-One

Words and Music by
Allen Cohen and Steven L. Rosenhaus

I've got the an - swer to the prob - lem
fine, though, 'cause here in - side my back - pack

Right here in my bag! It's the
I've got just the thing. It's the

Great Nor - thern All - in - One tool, So
Great Nor - thern All - in - One tool, So

228

ver-sa-tile and oh, so__ cool!__ It can gut a__ fish__ and
ver-sa-tile and oh, so__ cool!__ It can o-pen__ an - y

clean the__ scales,__ It can fix a flat,__ and then
can with__ ease,__ And it's e - ven great - er at

trim your__ nails!__ The best tool un - der the sun__
grat - ing__ cheese!__ The tool that's sec - ond to none__

Is the Great Nor-thern All - in - One!_
Is the Great Nor-thern All - in - One!_

234

There isn't much to say about the music of most comedy songs. Like many, this music tries to be light and pleasant, and not much more. As with any good theatrical comedy song, the music itself makes no attempt to be funny.

The music of the tag gave us some trouble. We found no way for it to end either fast or with the melody rising, as the first two refrains do. Instead we chose to give it a slower, "ad-lib" feeling, like the introductory verse, and to have the melody end low and dejectedly, with a transposition of the music for the first line of the refrain.

We might revise "The Great Northern All-in-One," but before deciding whether to keep it in the show we would first have it performed in front of an audience. More than any other type of song, a comedy song needs an audience. If they laugh out loud, you know that the song works. If they don't, you know that it needs to be either rewritten or replaced.

Writing a Charm Song

As discussed in chapter 5, the function of a charm song is to express something charming or likeable about the character singing, something that will get the audience to empathize with and root for the character. Chapter III of *The Prince and the Pauper* had seemed a natural spot for a charm song for both our protagonists: Edward brings Tom into his chamber, gives him food, and questions him about his daily life, which sounds as exotic to Edward as a prince's life does to Tom. Just as Tom envies the prince the luxury of his room and his clothes, Edward envies Tom his freedom to play and run barefoot in the mud. Obviously there is comedic material in this, and we considered writing a comedy song for this spot. But all the humor seemed on one side only: Edward envying Tom is funny, but Tom envying Edward is not. Since this might be the one duet in the show for the two boys, we wanted something that would be mutual and evenly distributed. And a single charm song for both would save time compared to separate songs for each.

As with "Poor Little Lad," we got a number of ideas from the novel, but we still had to do a considerable amount of invention. In the novel Edward questions Tom, but Tom doesn't question Edward. Perhaps a reader would assume that Tom doesn't need to ask about Edward's life because the common people already know a great deal about the daily life of the Prince of Wales, but this would not work on stage. Clearly we needed both Tom and Edward to sing something about themselves. This in turn suggested a structure. In the first A section, Edward would sing about his daily life, and Tom would react with envy. In the second, Tom would sing and Edward would be the envious one. We put Edward's account first because Tom's reaction is the expected one, while Edward's reaction to Tom's account is surprising and thus provides a humorous intensification or "build" to the number.

As discussed in chapter 3, in a musical most of the emotional high points are musicalized. The emotional crux of this scene is when the boys exchange clothes, and we wanted to incorporate that moment into the song. This suggested the remainder of the lyric structure: The third section would be a bridge in which both boys contemplated exchanging roles, if only for a moment. In the next A section the boys would alternate lines about what they might do in the other's shoes. Then would come an instrumental interlude, using music of the B section, during which the boys would exchange clothes. We imagined that the prince might have a double-paneled folding screen in his room, behind which the boys would change. Finally they would step out, each in the other's clothes, and sing a short tag to the effect that, as brief as this adventure might be, it would give them memories to last a lifetime.

We started searching for a title that would work for both boys. Our first ideas were along the lines of "That's the Life for Me" or "What I Wouldn't Give." Finally we chose a phrase from the novel, "Just Once." We were reluctant at first because these two syllables are full of sibilants and consonants, and because "once" is not a good syllable to sustain, nor does it have any useful rhymes; but we decided it was definitely the best central idea for the song. We then decided to structure each section with an eight-line stanza followed by a short tag like a refrain; the title phrase "just once" would occur at the beginning rather than the end of each tag, except at the very end of the song.

Next we selected lines from the things Tom says about himself in the novel, and added new ones. We came up with activities for Edward from elsewhere in the novel, from research, and from our imagination. We also had to leave out several wonderful lines from the novel because they would require far too much explanation.

Here are the first two stanzas of the lyric for "Just Once."

EDWARD

EACH DAY WHEN I AWAKE, MY SERVANTS DRESS ME
IN CUSTOMARY VELVET, SILK, AND SATIN.
I BREAK MY FAST WITH QUAIL AND GROUSE AND
   PHEASANT PIES,
AND THEN I STUDY FRENCH AND GREEK AND LATIN.

I EAT THE MIDDAY MEAL, AND THEN I PRACTICE ARTS OF
   WAR—
THE DAGGER, SWORD, AND OTHER WAYS TO FIGHT.
AND THEN THEY TEACH ME HISTORY AND ETIQUETTE FOR
   HOURS—

TOM

IT SOUNDS LIKE A PERPETUAL DELIGHT!
JUST ONCE, TO STUDY ALL I WANT AND HAVE ENOUGH TO
   EAT—
I'D GIVE MY ONLY SHIRT FOR SUCH A TREAT!

EDWARD

Tell me more about yourself. What is your day like?

TOM

MY FATHER WAKES ME ROUGHLY EVERY MORNING,
AND SENDS ME ON THE STREETS TO BEG A SHILLING.
BUT AFTERWARD I MEET MY FRIENDS AND WE
   PLAY GAMES—
AND OH, YOUR HIGHNESS, GAMES CAN BE QUITE THRILLING!
IN SUMMER WE GO SWIMMING IN THE RIVER AND THE
   BROOK,
OR DIVING IN AND RACING BACK TO SHORE.
FROM SAND WE FASHION CASTLES, OR MAKE PASTRIES OUT
   OF MUD—

EDWARD

I'VE NEVER HEARD SUCH MARVELS TOLD BEFORE!
JUST ONCE, TO FROLIC BAREFOOT IN THE MUD AS OTHERS
   DO—
IT WOULD BE A FANTASTIC DREAM COME TRUE!

This is as far as we took the lyric. In the next stanza, the bridge, the boys
would sing about this chance to have a marvelous adventure, a once-in-a-
lifetime opportunity to feel how the other half lives.

In the next A section the boys would become increasingly excited as they
alternated lines, fantasizing about leaving behind the unpleasant aspects of
their lives. For Tom this would include his father's beatings, his constant

hunger, and his ignorance; for Edward, the regimentation of his every waking hour, his virtual imprisonment in the castle, and his inability to ever be simply a ten-year-old boy. They would also imagine living as the other does: For Tom, this might involve wearing luxurious clothes and conversing with kings and lords; for Edward, being free to go anywhere unrecognized.

Then would come the interlude during which they exchange clothes. There could be some comic byplay with the boys tossing clothes back and forth, and occasionally peeking a head out from behind the panels of the folding screen to exchange a word or two of dialogue. Then they would appear and sing the final tag, an extended version of the previous tags, which would go something like this:

EDWARD

JUST ONCE, LIKE EVERY OTHER BOY, TO FREELY RUN AND PLAY—

TOM

JUST ONCE, TO EAT A DINNER AND NOT WORRY HOW TO PAY—

BOTH

AT LEAST I SHALL REMEMBER THAT I LIVED ONE PERFECT DAY, JUST ONCE!

For the musical setting, we wanted something moderate and pleasant, like most charm songs. In older shows this is usually achieved with a fox-trot beat and swing rhythm, as in such songs as "Once in Love with Amy," "Getting to Know You," and "Wouldn't It Be Loverly?" The fox-trot beat, however, sounds very dated now, and we had already used swing rhythm in "The Great Northern All-in-One," so we found a different beat and used a "straight eighth-note" (that is, not swung) rhythm. As with "Poor Little Lad," we tried to incorporate some modal sounds to give a slight period flavor; the Lydian mode with its raised fourth can be heard in the verse, and the Mixolydian mode with its lowered seventh at some of the cadences.

Just Once

Words and Music by
Allen Cohen and Steven L. Rosenhaus

As mentioned in chapter 6, the range of a song for children must usually be narrower than that of a song for adults. The first two stanzas of "Just Once" have a melodic range of an octave, which is very safe. In the remainder of the song we might allow the melody to extend one whole step higher, making an overall range of a ninth, which is still a safe range for children.

However, we feel a little uncertain about the musical setting of "Just Once." It is light and pleasant, seems appropriate for its ten-year-old singers, and has a nice flavor that does not sound excessively derivative. But while there is nothing particularly wrong with the melody, and it fits both the meter and meaning of the lyrics reasonably well, it is not very distinctive. With its eighth-note rhythm, repetitions, and sequences, it comes close to sounding like a "patter song"—the kind of syllable-crammed tune found in Gilbert and Sullivan operas, in which almost every bar is crammed with notes, and the melody is only one step above recitative. If we decided to keep this melody, it would be crucial not to allow the song to be played faster than the tempo indicated, so that it feels relaxed and moderate rather than quick and "pattery."

Writing a Rhythm Song

The clearest place we found for a rhythm song in *Life on Another Planet* was in Act One, Scene Nine, when Kari has decided to run away from both of her parents into the desert. This song would allow her to vent all the unhappiness, resentment, and rage she has been bottling up inside. Feeling angry, confused, and unable to deal with the pressure or to decide between her parents and their homes, Kari decides to break free. Our idea is that as soon as she finishes her phone call to Jen, she starts to write an e-mail on David's computer, addressed to both her parents, and sings to us as she is writing. Of course, she does not have to be sending an e-mail; she could simply hang up the phone and sing it. She could even sing it to Jen, but we felt it would be stronger if it were directed at her parents.

Her state of mind, and the fact that this scene would necessarily occur at the beginning of July, suggested a thematically appropriate date: the Fourth of July, or Independence Day. Kari would see it as her Independence Day as well.

The basic concept clearly suggested a structure. The first A section would be directed at Kari's father, the second at her mother. The last would be for both parents. In the bridge she would speculate about where she might go next, and then decide that, uncertain as running away is, it has to be better than staying with either Terri or David.

The next step was to decide what Kari would say in each stanza. Since in this scene she is very angry, and since David sees himself as being closer and more attentive to her than Terri is, it seemed natural for Kari to attack him on that front. As for Terri, we thought that Kari would challenge her complacent assumption that she and Kari are best buddies. For the last A section, Kari could vent her rage at both parents for never having given her a stable home environment and for making her choose between them now, and could end by rejecting both of them.

There seemed to be two prime candidates for the title, "Independence Day" and "The Fourth of July." At first we were inclined not to choose "The Fourth of July" for two reasons: As a title it did not identify the mood as clearly as "Independence Day," and—if we were writing an actual show—we would be concerned that non-American audiences might not know the meaning of the date. A disadvantage of "Independence Day" was that it had been the title of a popular song by Bruce Springsteen, but that was less of a problem. The Springsteen song is decades old and in a different genre, and in any case the title is too general to be solely identified with one particular song. Ultimately, however, we decided we preferred "The Fourth of July" precisely because it did not hit the nail on the head as baldly as "Independence Day." As for the international question, we knew that many foreigners have heard of the holiday, and the context of the song would make clear to everyone that the date was our Independence Day.

One idea we tried was to incorporate quotations from the Declaration of Independence into the lyrics. We thought there might be a way to use words or phrases like "in the course of human events," "self-evident," "inalienable rights," or even "A Declaration." Ultimately we found no way to make the quotations fit the context, and ended up using only the phrase "declare my independence." We also liked the idea of mentioning things associated with the July 4th holiday, such as fireworks and picnics. Although we didn't use these in the first two A sections, we might well put a fireworks metaphor into the final one. The first and third lines of the final section would end with "folks" instead of "Dad" or "Mom," and the last rhyme with the title line would be the word "goodbye."

Kari is a contemporary teenager, and because her anger in the scene is so strong, we knew we wanted this to have the feel of a rock song. That suggested a verse-refrain structure for each A section. This doesn't mean that every song for Kari would have this form, as "The Great Northern All-in-One" already did, but two songs with similar structure in a show, separated by an act break, are not too many.

Here are the first two stanzas of our lyric for "The Fourth of July."

KARI

(typing into the computer)

Dear . . . David . . .

YOU THINK YOU SEE RIGHT THROUGH ME, DAD.
YOU THINK YOU KNOW WHAT I'M FEELING.
WELL, HERE'S A LITTLE NEWS FLASH, DAD—
IT OUGHT TO MAKE YOU HIT THE CEILING!
YOUR LITTLE GIRL IS LEAVING YOU TODAY,
'CAUSE I'LL GO NUTS IF I DON'T GET AWAY.

IT'S THE FOURTH OF JULY,
THE DAY I DECLARE MY INDEPENDENCE.
IT'S THE FOURTH OF JULY,
AND TIME TO SET MYSELF FREE.
AND I HOPE YOU'LL COME TO SEE
IT WAS TIME FOR ME TO FLY . . .
'CAUSE IT'S THE FOURTH OF JULY!

And now for you, Terri . . .

YOU THINK WE'RE SUCH GOOD BUDDIES, MOM,
WITH ALL YOUR HUGGING AND KISSING.
WE'LL SEE HOW LONG IT TAKES YOU, MOM,
BEFORE YOU EVEN KNOW I'M MISSING!
YOUR LITTLE PAL WON'T BE THERE ANY MORE.
FIND SOMEONE ELSE TO SMOTHER, THEN IGNORE!

IT'S THE FOURTH OF JULY,
THE DAY I DECLARE MY INDEPENDENCE.
IT'S THE FOURTH OF JULY—
I NEED TO GO IT ALONE.
CAN I MAKE IT ON MY OWN?
WELL, I WON'T KNOW TILL I TRY . . .
AND IT'S THE FOURTH OF JULY!

Having previously written "The Great Northern All-in-One," we already had Kari's vocal type in mind. Considering the type of anger that the song needed to express, we knew that it needed at least a flavor of classic hard rock. Since an eighteen-year old girl is supposed to be able to sing it live, eight times a week, we also needed to ensure that the range of the song was not too wide, nor the tessitura too high.

The Fourth of July

Words and Music by
Allen Cohen and Steven L. Rosenhaus

Up-tempo Rock

You think you see_ right through_ me, Dad._ You think you know_ what I'm

think we're such_ good bud - dies, Mom,_ With all your hug - ging and

(sim.)

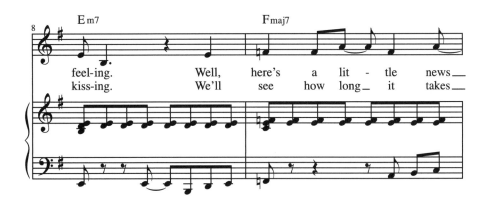

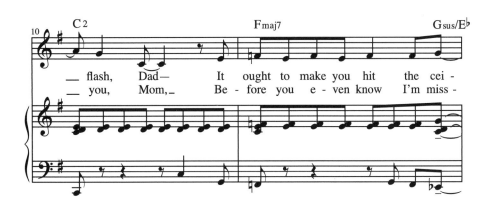

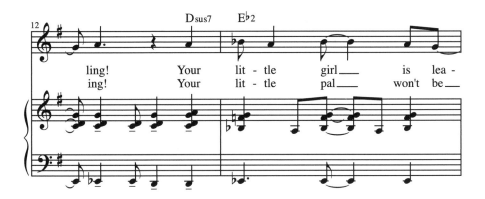

We knew we wanted the refrains to be in major. At first we thought of begining the verses in the relative minor, but we avoided that because it would make the key relations of this song too similar to those of "The Great Northern All-in-One." We did sneak in a temporary minor reference in measure 8, however.

We feel satisfied that the song has the flavor of rock, and that despite this the melody has a distinctive contour and the refrain is reasonably "catchy." The rock flavor also accounts for the blues-scale flatted third (the B-flat) in measure 22.

The melodic range of the first two sections is only an octave. In the bridge or the last section, the melody might go up to a C or D, making a tenth overall, which is not too wide a range. Depending on an actual singer's range, the song might need to be transposed down by a half step or whole step, but that is a small change that would cause no problems.

The Rest of the Show

This chapter has shown how you can begin the score for your own show. Because we chose only two songs for each project, some of the problems involved in writing an entire show did not arise for us. But this book gives you some of the tools you need to solve such problems.

You may be curious about the way the two of us collaborated in writing these songs and scenes. Although we have both written songs alone and in collaboration with others, these were the first songs we ever wrote with each other. They are also the first musical theater songs in which either one of us worked with a partner on both lyrics and music. These two factors made the collaboration both exciting and difficult. Undoubtedly with time the process would become easier and the results more inspired. Nevertheless, we are glad to have done it. Collaborating on both elements of a song turned out to be not as difficult as we feared. Writing lyrics with a partner is similar to writing lyrics or music by oneself, except that instead of tossing ideas back and forth within one's own head, one tosses them back and forth with another person. Even writing music together was not as strange as we thought it would be; one of us would sing or play an idea for a tune or a harmonic progression, the other would give his reaction, and together we would try to arrive at something we both liked. In the process we created songs that were both similar to and different from songs we would have written individually; our different backgrounds and aesthetic predilections required compromises, but the collaboration also took each of us in directions that we might never have tried on our own. Appendix C includes detailed discussions of collaboration and how you can find, and keep, the right collaborators.

None of the songs in this chapter is complete or finished in the way that theater songs ultimately need to be. There is one last crucial stage that they would need to go through before we could consider them finished. The next chapter explains that final stage.

CHAPTER 9

REWRITING

When you have finished writing the score and the libretto of your show, your work has only begun. What happens next is as important as the writing itself. In musical theater, *writing is rewriting*.

That sentence is so important that we're going to repeat it: *Writing is rewriting*. No matter how talented and experienced a writer you may be, that is not enough to make a successful show. If it were, every show written by Rodgers and Hammerstein, Lerner and Loewe, Bock and Harnick, Leonard Bernstein, or Stephen Sondheim would have been successful.

There are two stages of rewriting, and each has its own rationale. The first stage is the rewriting that you do after you have finished a song, a scene, or an entire show. Before you show it to anyone else, wait a few days, then go back and take another look at it. This takes the ability to look coldly and objectively at your own work, to spot flaws and be willing to do whatever it takes to fix them. The rationale behind this stage of rewriting is that there is no such thing as a perfect theater song or scene. Any song or scene can be improved.

When you are collaborating, the task of rewriting is more delicate. But it is essential that collaborators be able to critique each other's work, and that each writer be able to hear criticism from collaborators without taking it personally. As long as the criticism is constructive—and the goal behind it is to produce the best show possible, rather than to attack the other person—there should be no offense given or taken.

The second stage of rewriting is the most important of all. This is the writing that you do after you have presented a song, a scene, or an entire show to an audience.

In musical theater the audience's reaction may be more important than in any other art form. Theater truly exists only in live performance. And musical theater is a collaborative, heterogeneous, commercial medium. Not even the most talented and brilliant practitioners have been able to predict how an audience will react when all the elements are put together and presented in a live performance.

The audience does not have to be fifteen hundred people in a large theater. It could be a group of acquaintances in a living room, or a group of peers in a songwriting class or workshop. It could be a group of strangers, during an "open mike" night at a club somewhere. It could be at a reading of the show; or at a "staged reading," in which the actors may walk around but still hold their scripts; or at a low-budget showcase performance. If there are enough people, the exact nature of the audience doesn't really matter. They are an audience, and they will tell you whether your song or your scene is finished and ready. And whatever their response, you must listen to them.

How many people are enough to make an audience? Obviously, the more the better. In our experience, half a dozen is too small; we would say that a dozen is the bare minimum. The larger the audience, the more accurate its reaction will be as a gauge of how other audiences will react. If fifteen people in a room laugh at your comedy song, the odds are very good that one hundred and fifty will. And if one hundred and fifty people in a small theater laugh at your comedy song, then you can be pretty certain that fifteen hundred people in a large theater will as well.

This is the reason why our sample songs in the previous chapter are nothing more than tutorials. If we were really writing either of our sample shows, we would need to hear each song performed. We would not consider a song finished until we heard it performed in the context of the entire show in front of an audience. Only then would we know whether these songs truly worked well enough to be considered finished.

Peter Stone, author of the superb libretto for *1776*, once wrote an article saying that if you ask any individual member of an audience for a critique of your work, that person will probably be wrong—but collectively the audience is always right. If the audience doesn't laugh at your comedy song, if it isn't moved by your ballad or your tender scene, then something is wrong with the song or scene. "Throw it out, change it, realign it, do something, because they are rejecting it . . . and to ignore it is absolutely a death wish."

As Stone says, that doesn't mean you must throw it out. Perhaps a small change in the lyric or dialogue will fix it. Perhaps something in the lead-in makes the song feel redundant; perhaps the accompaniment is too loud, or the tempo too fast, for the audience to understand the words; perhaps some crucial element needed to set up a scene is missing earlier in the show. If there is any worth to the song or the scene, it can theoretically be fixed, except in one case: When directors or choreographers stage it in a way that does not let it come across as you believe it needs to, all you can do is try to change their minds. In a professional production, however, the problem is usually with the material rather than with the performance.

Writing a musical is an act of creation. But the authors alone are not enough to bring a musical into existence. More than any other art form, except perhaps film, musical theater is a collaborative medium that requires the talents and the ideas of many people in different functions to make it come to life. Without directors, choreographers, designers, arrangers, actors, and musicians, a written musical cannot be turned into a produced musical. Without a production, it does not exist as a piece of theater except in potential. This means that, to the extent you can, you must try to choose your interpretive collaborators as carefully as you choose your writing collaborators. It also means that the writing of the show is not finished until it has been performed, rewritten, and performed and rewritten again and again. We hope that this book helps you write your show and bring it to that final stage—and beyond.

Appendix A

A Brief Historical Overview

Beginnings

The antecedents of contemporary American musical theater include such distant fore-bears as Greek tragedy, which incorporated song and dance, and eighteenth-century *opera buffa* such as *The Marriage of Figaro* by Mozart and da Ponte. More recent ancestors include vaudeville and the minstrel show. Musical theater's date of birth is usually given somewhat arbitrarily as 1866, the year in which *The Black Crook* opened. This was a fortuitous and bizarre amalgam of two theatrical productions: a French ballet troupe that found itself stranded in New York when the theater in which it was to appear burned down, and an American melodrama freely plagiarized from the story of the opera *Der Freischütz* by von Weber and Kind. The two productions were combined into one, with spectacular sets and costumes, songs (including "You Naughty Naughty Men," which became a hit), and—not least important—French ballerinas in pink tights showing more of their legs than American theater audiences had ever seen. The show was a smash success and quickly spawned many imitations.

Thus began what could be called the infancy of American musical theater (approximately 1866 to 1904). It was dominated by two trends: imitations of European operettas or comic operas by Offenbach, Johann Strauss, or Gilbert and Sullivan; and musical comedies that were more American in style, in subject matter, and in their broad humor, such as the *Mulligan Guards* series by Edward Harrigan and David Braham (called the "Harrigan and Hart" shows, after their stars) or the 1891 hit *A Trip to Chinatown*, which included such tunes as "The Bowery" and "Reuben and Cynthia." The librettos of this period are full of insulting ethnic stereotypes, silly and convoluted plots, excruciatingly bad puns that furnish much of the comedy, and pasteboard characters. Almost without exception, the songs have lost whatever interest they once had.

The childhood of musical theater could be said to have begun around 1904, with the musicals of George M. Cohan. While his use of slang and colloquial language was not new, his energy and rapid-fire pacing established new standards for the field, and his combination of brash optimism and sentiment seemed to embody the American spirit.

His shows have long since disappeared, but even one hundred years later, Cohan tunes such as "Give My Regards to Broadway," "Yankee Doodle Dandy," and "You're a Grand Old Flag" still resonate within the memories of most Americans.

The other main trend in musical theater in those early days was the revue, the heyday of which lasted until the Great Depression of the 1930s. In the first three decades of the twentieth century a host of revues appeared on Broadway, including the Ziegfeld *Follies*, the George White *Scandals*, the Shubert *Passing Shows*, and the Earl Carroll *Vanities*, each of which featured a wide variety of performers and acts, dozens of chorus girls, and songs by some of the most prominent writers of the day. Irving Berlin got his start in the *Follies*, and George Gershwin later achieved prominence with his songs in the *Scandals*.

The Song Era

The impetus for the next period in the history of musical theater is usually credited to the "Princess Theater musicals," which were written primarily by lyricists and librettists Guy Bolton and P. G. Wodehouse and composer Jerome Kern in the late 1910s and early 1920s. They are often said to have ushered in a new era of sophisticated humor, songs of high quality, and well-integrated librettos. In fact, the Princess shows differ from the other musicals of their era only slightly. Their real distinction resulted from being essentially off-Broadway shows, with low budgets, few sets, small orchestras, and smaller casts than usual. Their plots remained silly, the dialogue and humor strained and artificial, and the songs, despite some gems, no better than those of many contemporaneous shows.

But if the Princess Theater era did not usher in a golden age of musical theater, it coincided with a golden age of theatrical song. Kern's songs got better and better, while Irving Berlin appeared upon the Broadway scene, followed shortly by the Gershwins, Rodgers and Hart, Vincent Youmans, Cole Porter, and De Sylva, Brown, and Henderson, and later by Schwartz and Dietz, Harold Arlen, E. Y. Harburg, and Vernon Duke. In this era appeared many of the songs that became the "standards" of jazz and the core repertoire of popular music in the first half of the twentieth century. While musical theater had not yet reached maturity, this period (about 1915 to 1943) could be considered its adolescence.

Virtually no shows from this period are still performed today. Besides *Porgy and Bess*, the opera by the Gershwins and DuBose Heyward (1935), the only important exception is *Show Boat* by Oscar Hammerstein II and Jerome Kern (1927). *Show Boat* was a towering achievement in many ways. Its score was of extremely high quality; its libretto was ambitious in scope and bold in its treatment of racism and miscegenation. It also had many flaws and many remnants of old musical comedy and operetta, but it came closer to a true integration of libretto and score than had any show that preceded it— or any show that followed it for sixteen years. Despite its success, *Show Boat* had little influence at the time. Neither audiences nor critics expected the libretto of a musical to be anything but serviceable, and they usually got what they expected.

The "Golden Age"

The most significant event in musical theater history occurred on March 31, 1943, when Richard Rodgers and Oscar Hammerstein II, with the help of director Rouben

Mamoulian and choreographer Agnes de Mille, brought *Oklahoma!* to Broadway. Although its plot was essentially a trifle about whether a nice cowboy or a surly farmhand would take a girl to a dance, *Oklahoma!* was a giant step forward in the quality and integration of musical theater. It broke with conventions and experimented with new techniques, most notably the Freudian "dream ballet" that ended the first act. It also showed surprising realism in dealing with the dark underside of its sunny story, and it included the death of a major character. Both the libretto and the score were unequalled by any that had yet appeared. To a much greater extent than *Show Boat*, *Oklahoma!* combined the lyricism and seriousness of operetta with the American pizzazz and colloquial flavor of musical comedy, thus uniting the two streams into a single entity with the strengths of both. Most important, it was the first completely integrated work of musical theater. Every aspect of the writing and the production furthered and enhanced the drama.

In each of these aspects *Oklahoma!* was so successful and influential that it can be said to have changed the course of the genre and ushered in the maturity, the "golden age," of musical theater. Almost all of the classic musicals appeared during the next twenty-one years. (See the list in Appendix D.) Rodgers and Hammerstein, Lerner and Loewe, Irving Berlin, Cole Porter, Frank Loesser, E. Y. Harburg, Leonard Bernstein, Comden and Green, Jule Styne, and Bock and Harnick were all at their peak—as were such librettists as Abe Burrows, Arthur Laurents, Joseph Stein, Burt Shevelove, and Larry Gelbart. Like *Oklahoma!*, the best shows of this period have not only songs but also librettos of high quality, and a high level of integration of scene, song, and movement.

Traditional directors such as George Abbott, George S. Kaufman, Josh Logan, Moss Hart, and Abe Burrows were at the height of their artistic powers during this period, which was also the golden age of theater dance. Numbers that were not only spectacular but integral to the shows in which they appeared were created by choreographers such as Jack Cole (*Kismet*), Michael Kidd (*Guys and Dolls*), Onna White (*The Music Man*), Gower Champion (*Bye Bye Birdie* and *Hello, Dolly!*), Bob Fosse at his least self-indulgent (*Pajama Game, Damn Yankees*, and *How to Succeed in Business without Really Trying*), and Jerome Robbins (*On the Town, The King and I, West Side Story, Gypsy*, and *Fiddler on the Roof*).

During this period Robbins, who is considered by many to be the one true genius of the musical theater, conceived and directed two shows that represent pinnacles of musical theater achievement: *West Side Story*, with its seamless blend of drama, song, and dance, and *Fiddler on the Roof*, the last and perhaps greatest show of the golden age, the paragon of the traditional book musical.

Rock and the Concept Musical

After *Fiddler on the Roof* in 1964, the golden age came to an end, and musical theater entered a phase that, in keeping with our metaphor, could be called old age. (This phase has now lasted longer than the golden age.) There is no single cause, but a number of contributing factors. For one, many of the best traditional writers either died or found themselves unable to recapture the magic of their previous work. The same was true of many of the best directors and choreographers—except for Robbins, who left musical theater to devote himself full-time to ballet. Equally significant were the geometrical increase in the costs of producing on Broadway and the continual decline in the numbers of new musicals produced each year. In the 1950s many New York nightclub and cabaret revues had helped fill the gap left by the shrinking opportunities on Broadway; in these intimate revues, young songwriters had been able to polish their craft one song at a time. During the 1960s these venues disappeared. As a result, new writers had fewer and fewer places in which to experiment, to fail, and to learn, before getting what might be their one chance at the "big time."

In addition, the Beatles came to America in 1964, and over the rest of the decade rock music grew from a teenage fad to the universal language of popular music. There was a deep divide between the popular styles with which the great songwriters had grown up and the new popular styles, and none of these songwriters was willing or able to cross it; neither they nor their younger counterparts such as Strouse and Adams, Jerry Herman, Jones and Schmidt, Kander and Ebb, and Stephen Sondheim ever learned to rock.

Rock began to appear in musical theater, first as an object of satire, then in some abortive experiments, and finally with tremendous success in *Hair* in 1968, although much of *Hair's* success was due to its brief nude scene. This paved the way for other rock shows that were concerts as much as they were plays, such as *Jesus Christ Superstar*. Because of the conservative nature of Broadway audiences, and because undiluted rock music is inherently undramatic, rock styles have never achieved the universal hegemony in musical theater that they have in popular music. Younger theater composers have assimilated rock or pop styles into their musical vocabulary, but even today there is a wide variety of musical styles, from shows that sound completely uninfluenced by rock to those that sound like rock concerts. Many shows still mix old and new styles, sometimes uncomfortably.

In traditional musical theater two divergent trends emerged. In one, following Rodgers and Hammerstein, dance gradually became less important or disappeared, as in the later shows directed by Harold Prince. In the other trend, following Jerome Robbins and Gower Champion, dance and movement became even more important in the hands of director-choreographers like Bob Fosse and Michael Bennett. The 1971 production of *Follies*, co-directed by Prince and Bennett, can be seen as a pivotal event in this divergence; after *Follies*, Prince directed few shows in which dance had a significant role, and Bennett took complete control of his shows as director-choreographer.

Another trend in this period was the concept musical, especially in the shows directed by Harold Prince, such as *Cabaret* and *Company*. An important partnership between director Prince and songwriter Stephen Sondheim began with *Company* in 1970 and lasted for over ten years, during which Sondheim wrote many of the best songs in musical theater history. All of their shows from this period—*Company, Follies, A Little Night Music, Pacific Overtures, Sweeney Todd*, and *Merrily We Roll Along*—are concept musicals to a greater or lesser degree, and all show daring experimentation and exciting innovation, both technically and in subject matter. They represent most of the peaks of musical theater since 1964, although they fail to match the greatest golden age musicals in portraying or creating strong emotional involvement. The greatest example of the concept musical, perhaps because it took the most liberties with its concept, was *A Chorus Line*, which was directed and choreographed by Michael Bennett in 1975. As of this writing it remains the longest-running musical by Americans in musical theater history, and no Broadway musical since has matched its quality.

Pop Opera

A second "British invasion" that proved as significant as the arrival of the Beatles began in 1980 with *Evita*, the second musical by Tim Rice and Andrew Lloyd Webber to come to America and become a hit (the first being *Jesus Christ Superstar*). *Evita* was followed by a succession of Lloyd Webber musicals: *Joseph and the Amazing Technicolor Dreamcoat, Cats, Song and Dance, Starlight Express, Phantom of the Opera, Aspects of Love,* and *Sunset Boulevard*. Most of these through-composed pop operas were commercially (though not critically) successful, and *Cats* and *Phantom of the Opera* became two of the most successful shows in history. Although many American musical theater professionals find them mediocre at best, neither their popular success nor their influence can be denied. London has seen many imitations of Webber and Rice musicals, and Paris has produced its own: *Les Misérables* by Alain Boublil and Claude-Michel Schönberg. Imitating *Evita* in structure and technique, if not in craft, it became almost as successful as *Cats*, and Boublil and Schönberg followed it with *Miss Saigon*. It was inevitable that American imitations of Lloyd Webber shows would appear, and eventually one did, when *Jekyll and Hyde* appeared on Broadway in 1997.

None of these poperettas contains real rock music, although *Superstar* comes the closest. True rock music only made it to a Broadway book show with the appearance in 1993 of *The Who's Tommy*, twenty-five years after that rock opera appeared on disc and started the trend. Besides *Tommy* the closest thing to rock in a musical, in terms of both composition and performance style, has been the score of *Rent*, the phenomenally successful 1996 adaptation of *La Bohème*. Mention must also be made of the off-Broadway hit *Little Shop of Horrors*, which brilliantly used the sounds of early-1960s rock 'n' roll to evoke its campy period ambience.

The Present Outlook

It is hard to discern current trends in musical theater, and even harder to predict what might happen next. In the recent past, one interesting trend seems to be the gradual disappearance of the pop operas of Andrew Lloyd Webber and his imitators. Lloyd Webber's *Sunset Boulevard* was not very successful—somewhat ironically, because it included some of Webber's best songwriting and was closer to a traditional book show than any of his hits. His next show never made it to Broadway, and as of 2005 he had not had another success in over fifteen years. Boublil and Schönberg have not had a success since *Miss Saigon*, and Frank Wildhorn, the composer of the mildly successful *Jekyll and Hyde*, followed it with three failures.

One Broadway trend that has become virtually an epidemic is the frequency of revivals of older shows. To some extent this may seem inevitable, given the falloff in quality of new shows in the past decades; but it also reflects changes in the economics of Broadway, as well as the disappearance of the old-time Broadway producers, who were willing to take chances on new shows without having to see them produced elsewhere first.

Another apparent trend is an increasingly heavy reliance on recycling movies of such recent vintage that they have had no time to lie fallow and acquire new meanings or perspectives for an audience. The Walt Disney Company began turning cartoons into expensive Broadway productions with *Beauty and the Beast* in 1994, when the movie was only three years old. In comparison, the stage version of *The Producers* seemed almost classically restrained in adapting a movie more than thirty years old. *New York Times* critic Ben Brantley has referred to such musicals, including the more recent *Spamalot* (based on *Monty Python and the Holy Grail*) and *Mamma Mia!* (built around the songs of ABBA), as belonging to a new genre he calls "scrapbook musical theater," which repackages beloved popular movies or songs, and thus presumably guarantees an audience who will pay a hundred dollars each for a trip down Memory Lane. Another type of recycling can be seen in the recent epidemic of self-referential numbers and shows that parody other shows, such as *Urinetown, The Producers*, and *Spamalot*. Parody can be very entertaining, but for it to become the chief source of musical comedy is a sign of creative impoverishment.

If musical theater is to escape the final stage of mummification, what it needs more than ever are high standards of craftsmanship and a willingness to try the new—new original ideas, new and fresh perspectives in adapting older works—while retaining the best of the old traditions. Of all those traditions, none is more important than the premise with which we started this book: that the essence of musical theater is the representation and evocation of emotion through the union of drama and music. The flashiest and most expensive technology—which in any case cannot match that of film—is no replacement for characters that we care about and stories that touch our hearts through song and dance. With high quality of craftsmanship and emotionally compelling stories, the musical can again hold the central position in our cultural life that it once did. We hope that this book may encourage and help new writers to lead the way.

TOOLS

For Everyone

Since actors are the performers and interpreters of dramatic characters, an understanding of how they prepare and perform their roles is invaluable in learning how to write effectively for them. All of the following are considered standard textbooks.

Stanislavsky, Konstantin. *An Actor Prepares.*
Boleslavsky, Richard. *Acting: The First Six Lessons.*
Rockwood, Jerome. *The Craftsmen of Dionysus.*
Hagen, Uta. *A Challenge for the Actor.*
Meisner, Sanford. *Meisner on Acting.*
Bruder, Melissa, et al. *A Practical Handbook for the Actor.*

The author of the following books was a longtime Broadway conductor, a brilliant analyst and teacher who created the BMI Musical Theater Workshop and ran it for fifteen years. Although they are now slightly out of date, these books, and in particular *Words with Music*, contain essential discussions of the aesthetics of musical theater and many valuable insights into its creation.

Engel, Lehman. *The American Musical Theater: A Consideration.*
Engel, Lehman. *The Making of a Musical.*
Engel, Lehman. *Words with Music.*

The following two books deal with practicalities and legalities of production. The first is often revised to remain up-to-date. The second was written by an experienced lawyer.

Theatre Communications Group. *Dramatists Sourcebook: Complete Opportunities for Playwrights, Translators, Composers, Lyricists and Librettists.*
Singer, Dana. *Stage Writers Handbook: A Complete Business Guide for Playwrights, Composers, Lyricists and Librettists.*

The following books are memoirs by writers or directors, collections of lyrics, and chronicles of particular shows, all of which offer object lessons or inspiration for writers. The Goldman book may be the best ever written about the Broadway theater: witty, provocative, and full of fascinating insights. Even though it is more than thirty years old, it remains all too relevant. Hammerstein's introduction to his collection is a wonderful essay on writing musicals and lyrics in particular. The two Guernsey books, collected from issues of the *Dramatists Guild Quarterly*, include many valuable articles and opinions by veteran writers; *Playwrights Lyricists Composers* contains Stephen Sondheim's talk on "Theater Lyrics," probably the best essay on the subject to have appeared in print.

Altman, Richard. *The Making of a Musical: Fiddler on the Roof.*
Berlin, Irving. *The Complete Lyrics of Irving Berlin.*
Bryer, Jackson R., and Richard A. Davison, eds. *The Art of the American Musical.*
Davis, Christopher. *The Producer.*
Dunn, Don. *The Making of No No Nanette.*
Gershwin, Ira. *Lyrics on Several Occasions.*
Goldman, William. *The Season.*
Gottlieb, Robert, and Robert Kimball, editors. *Reading Lyrics.*
Grant, Mark N. *The Rise and Fall of the Broadway Musical.*
Guernsey, Otis L., Jr., editor. *Broadway Song and Story.*
Guernsey, Otis L., Jr., editor. *Playwrights Lyricists Composers on Theater.*
Hammerstein, Oscar, II. *Lyrics.*
Hart, Lorenz. *The Complete Lyrics of Lorenz Hart.*
Jones, Tom. *Making Musicals.*
Kander, John, and Fred Ebb. *Colored Lights.*
Lerner, Alan Jay. *The Street Where I Live.*
Loesser, Frank. *The Complete Lyrics of Frank Loesser.*
Logan, Joshua. *Josh.*
Porter, Cole. *The Complete Lyrics of Cole Porter.*

Anyone seriously intending to pursue a career as a theatrical writer should apply for membership in the Dramatists Guild. It offers many informational services to members, and helps to protect their rights in contracts.

For Librettists

The first necessity for anyone writing in English is the biggest and best unabridged dictionary available. The second necessity is a good thesaurus, such as one of the many *Roget's*. A book of popular quotations, such as *Bartlett's*, and dictionaries of American and other slang are also valuable, as are dictionaries for foreign languages. There are online equivalents for all of these books, but none is as good.

The following are recommended books on dramatic writing. Many playwrights and screenwriters consider the Egri book to be the best there is on the subject. The Hatcher is also very highly regarded. The other two books deal primarily with screenplays, but most of the principles that they discuss apply to musical theater as well. Although its analyses are extremely schematic, the McKee has many valuable insights into the aesthetics of drama and the problems of adaptation.

Egri, Lajos. *The Art of Dramatic Writing.*
Field, Syd. *Screenplay: The Foundations of Screenwriting.*
Hatcher, Jeffrey. *The Art and Craft of Playwriting.*
McKee, Robert. *Story.*

For Lyricists

In addition to all the books already mentioned, the most important reference book for a lyricist is a good rhyming dictionary. There are many, but most writers consider the best to be the one edited by Clement Wood, now called *The Complete Rhyming Dictionary Revised*, edited by Clement Wood and revised by Ronald Bogus. There are also software programs that perform the functions of a rhyming dictionary and the books mentioned in the previous section.

For Composers

The following are useful reference books. It may seem strange to include Arnold Schoenberg in this list, but his book is full of fruitful suggestions for composers on ways of developing and expanding motivic or thematic material.

Gerou, Tom, and Linda Lusk. *Essential Dictionary of Music Notation.*
Grout, Donald Jay, and Claude V. Palisca. *A History of Western Music.*
Palisca, Claude V., editor. *Norton Anthology of Western Music.*
Randel, Don. *The New Harvard Dictionary of Music.*
Schoenberg, Arnold, edited by Gerald Strang and Leonard Stein. *Fundamentals of Musical Composition.*
Vinci, Albert C. *Fundamentals of Traditional Musical Notation.*

Good books about music theory are ubiquitous and easily found, and there are many books on popular songwriting. But the two books following are among the very few about composition for musical theater. The Banfield book is an insightful scholarly study. The Horowitz book may not be as interesting as its title would suggest, but it does include some thought-provoking ideas, opinions, and information.

Banfield, Stephen. *Sondheim's Broadway Musicals.*
Horowitz, Mark Eden. *Sondheim on Music.*

It is also a good idea for any composer who intends to write vocal music—or for any lyricist—to take voice lessons with a good and experienced teacher, at least for a few years. Lessons are valuable, not only because there may be times when you have to present your own songs, but also because the better you understand how singers breathe and produce sound, the better you will be able to write for them.

LOGISTICS AND LEGALITIES

Collaboration

Most writers need collaborators. People who have the interest and ability, not to mention the time, to single-handedly write book, lyrics, and music for a show are extremely rare. Even those interested in doing everything themselves need to be able to look at their own work critically and decide whether they are truly capable of doing it all, and doing it well.

Finding collaborators, however, can be difficult, and may take time, creativity, and persistence. One way is through ads and notices in the theater "trade" papers such as *Back Stage* or *Show Business*, or in weekly neighborhood papers; in New York City this would include papers like the *Village Voice* or *New York Press*. Local bulletin boards can also be useful, especially for college students. And nowadays the Internet offers many other options. For instance, the Dramatists Guild website has an electronic bulletin board where members may search for writing partners; the ASCAP website has a "Collaborator Corner" open to anyone looking for partners, even if they are not ASCAP members; and a recent group called Musical Makers, dedicated to the development of new musical theater projects, has a website where members can make a "Collaboration Connection."

Let's say that you have passed the biggest hurdle, and found one or more potential collaborators. The next problem is finding the *right* collaborator, and making sure that your collaboration will work.

There are a number of requirements for a writing partnership to work. They include a comparable degree of experience and success; a comparable level of technical and artistic skill; similarity of tastes in theater and musical theater; complementary work habits; complementary social, business, and performing skills; the ability to give and take criticism in a constructive way; and an indefinable something that can be called "chemistry." Let's consider these one at a time.

The first requirement is a comparable degree of experience and success. Very rarely will a writer with a significant track record take a chance on working with one who is considerably less experienced. Sometimes an experienced songwriter will collaborate with someone with significant experience in another literary field, as when Harold Arlen collaborated with novelist Truman Capote on the score of *House of Flowers*, or when Kurt Weill collaborated with playwright Maxwell Anderson on *Knickerbocker Holiday* and *Lost in the Stars*. But very few Broadway shows have been collaborations between veterans and novices. The major exception is *Phantom of the Opera*, for which Andrew Lloyd Webber chose an unknown, Charles Hart, to write the lyrics. Certainly Hart benefited from this collaboration, but Lloyd Webber was clearly the boss of this team, and actually brought in an old collaborator of his to write some of the book and lyrics.

The second and third requirements—comparable levels of skill and similar theatrical tastes—are probably obvious enough to need no elaboration. But "complementary work habits" and "complementary social, business, and performing skills" need some explanation. It might seem that work habits between collaborators should be identical rather than complementary, and it certainly can't hurt for writing partners to have similar habits. But even a collaboration between a writer who keeps diligently to a rigid schedule and a writer who procrastinates for weeks and then delivers the goods in a burst of activity can be successful—as witness the twenty-year partnership of Rodgers and Hart. Any combination of working habits is viable, provided that the collaborators are all comfortable with it. The same is true for any division of social, business, and performing tasks among the collaborators. So long as at least one of them is able to deal with outsiders, make financial arrangements, or perform their songs, then it is unnecessary for them all to have the same abilities.

It is also essential that each writer be able to take criticism from collaborators without taking it personally. Even more essential is the ability to *give* criticism to collaborators, without giving offense. There have been a number of shows in which all the creative participants were talented and celebrated veterans—for example *Mr. President*, with a score by Irving Berlin, a book by Lindsay and Crouse, and direction by Joshua Logan—which failed, at least in part, because the creators were reluctant to criticize each other's work.

The final element necessary for a good collaboration can be called "chemistry." As with a romantic relationship, a writing partnership needs a certain indefinable something that allows each writer to enjoy the collaboration. It is not necessary that the collaborators actually like each other, so long as they can work together. The partnership of Gilbert and Sullivan is an example. (Of course, it might have lasted longer if it had been more friendly.) Another sign of a collaboration with good chemistry is that it stimulates all the partners to do their best, and produces a result that is more than the sum of its parts.

A general requirement for any collaboration is a period of preparation, of intense and detailed discussion about every aspect of the intended show—the theme, the plot, the style, every scene, every character, and ultimately every song. Without such preparation and discussion, a show can end up as a hodgepodge. Even if the quality of each individual contribution is first-class, there is a good chance that each collaborator will "write a different show," with different ideas about the theme, the spine of the show, or the relative significance of scenes or characters. Many shows have run aground on the shoals of such differences. Sheldon Harnick has likened this process to people laying railroad track from two different directions toward a central point, and discovering when they approach the center that their trackbeds will lie half a mile away from each other. His prescription for avoiding such a disaster is "probing discussions" about the theme, the style, and the plot among all collaborators. As Stephen Sondheim has said, "The hardest aspect of writing a musical is to be sure that you and your collaborators are writing the same show."

There is one famous exception to the above principle. Before *Guys and Dolls* went into production, it had a book by Jo Swerling. The producers decided that the book was not working, and they brought in Abe Burrows to write a new book. Since most of Frank Loesser's score had been written already, Burrows wrote his book around the already-existing songs. In this case the result was a Broadway classic; but as many recent "cover" shows have demonstrated, the odds of this approach being successful are extremely low. And it's worth pointing out that while Burrows's version of the book was new, both versions, as well as Loesser's score, were based on the same material: the same stories, the same style, and the same characters.

Once the collaborators are in agreement about the general features of the plot, the theme, and the spotting of the songs, they can start to work separately; the librettist can begin to flesh out the book scenes, while the songwriters can begin to work on the songs. Of course, the writers need to stay in constant communication with each other as they write. This allows for adjustments in the dialogue, lyrics, and music as the show is written and refined.

At this point may arise the age-old question of whether the lyrics or the music comes first. When Richard Rodgers wrote with Lorenz Hart for the first two decades of his career, the music was almost always written first. When Rodgers worked with Oscar Hammerstein II on *Oklahoma!* and for the next seventeen years, the lyrics were almost always written first. This difference in procedure was due not only to the differences between the two lyricists, but to the differences between the types of shows they were trying to write. Both methods, of course, were successful. But modern book musicals are more tightly constructed and integrated than Rodgers and Hart shows. Most contemporary musical theater writers feel that it is more logical for the lyrics to come first, because this enables each song to be better integrated with the rest of the show in terms of character and situation.

Nevertheless, it is true of many golden age Broadway songs that the music, or most of it, was written first. For instance, when Alan Jay Lerner and Frederick Loewe had discussed a new song in great detail, librettist-lyricist Lerner would come up with a title. Then with Lerner present, composer Loewe would begin to improvise tunes which seemed potentially fitting. When he played one that they both liked, Lerner would then go off and write the lyrics. That is how they wrote most of the scores for such shows as *Brigadoon* and *My Fair Lady*. This was also essentially the working method of Sheldon Harnick and Jerry Bock, who wrote the scores for shows like *Fiorello!* and *Fiddler on the Roof*. (For that matter, both Hammerstein and the team of Comden and Green habitually wrote lyrics to old melodies before giving them to their composing partners—so in a way they also worked to music written first.) Neither method is unquestionably superior. But it should be clear from the description of Lerner and Loewe's practice that whether lyrics precede music or vice versa, preliminary discussion remains the first and most essential step.

If one person is writing both music and lyrics, the give-and-take of adjustment and rewriting is much easier. For instance, Stephen Sondheim has said that while jotting down ideas for lyric phrases and rhymes he often tries to arrive at a title, and then to assemble an evocative accompaniment figure or "vamp," although he does not settle on a final melody until the lyric is finished. In its own way this can be seen to parallel Lerner and Loewe's procedure. Similarly, Irving Berlin said that he wrote words and music at the same time to ensure that they "fit."

Collaboration Agreements

Some collaborations have lasted for decades with nothing more formal than a hand-shake. But it's a sad fact that partnerships do break up, and we live in a litigious society. For that reason, you might consider drawing up a collaboration agreement with your associates. An agreement can serve to prevent later disputes over decision-making, money, writer credits, or other matters. Production contracts or publication agreements often help clarify many of these points, but they may not encompass everything. In addition, a collaboration agreement is made solely among the writers, while production or publication agreements almost always involve third parties.

If you decide to have a formal collaboration agreement, be sure it is written in straightforward, understandable language, and that it covers the points explained below. Your agreement may also include other issues specific to your situation.

The date of the agreement. This is the month, date, and year in which the agreement takes effect.

Who is involved. The agreement should list each of the writers and their address of record, and explain the capacity in which each is acting, as for example, "Joe Words (hereafter 'Librettist') of 123 South Main Street, Teaneck, NJ 00000," and so on.

The purpose and extent of the agreement. The agreement should specify that the collaborators will work together on a specific project, according to the terms explained elsewhere in the agreement. The project should be named, even if the name is changed later; each agreement should be limited in extent to a specific project.

After you establish that preliminary information, you will need to clarify a variety of issues, including ownership of the work, division of income and expenses, and how the collaboration is to work.

Ownership of the material. The writing team should agree on how to share ownership of the work. One possibility is ownership by equal shares among the writers, with a third each for the librettist, lyricist, and composer. You can also divide ownership in half between libretto and score, which would give the composer and lyricist ownership of the songs and the librettist ownership of the script. In case the work is later published, you may also agree to grant control of the work to the publisher.

Division of income. This will be based on the ownership of the material. Equally divided ownership results in an equally divided income stream. Ordinarily book, lyrics, and music are each considered a separate function. Thus when one person performs two functions—for instance, when the librettist is also the lyricist—that person receives two thirds of any income, and the composer receives one third. If on the other hand the book is considered one function and the songs another, then the librettist earns half of any income and the composer and lyricist share the remaining half. This may not be a fair distribution unless the composer is also the lyricist, in which case each of the two partners would earn half.

It should be pointed out that this agreement covers the distribution of all income, not just the royalties paid to the writers by a show that is running commercially. Income may also be generated from translations, the sale of original cast albums, the synchronization and other licensing rights for the use of the music in other media, or from publication. Unless you agree otherwise, only the composer and lyricist will receive income from use of the music outside the show and without the script. Similarly, the composer will not receive income from publication of the libretto, unless this is agreed to in advance.

Who makes decisions. Artistic questions often arise, and you have to decide how to handle them. You can choose to work by majority rule (if there are three or more collaborators), or by unanimous vote, or by any other means you find mutually satisfactory.

Expenses. This will determine whether expenses such as copying, postage, option payments, and so on are to be paid by the individuals who incur them or by the group as a whole.

Agents. Collaborators often use a single agent to represent them all. Whatever you decide about representation should be mentioned in the agreement.

Billing. Although it may seem petty, disputes over credits can ruin a good collaboration. You should decide whether the names are to be listed alphabetically, by the writer's function, or by some other means. You might also agree whether all collaborators' names will appear in the same typeface, the same type size, and so on. In case of a production you may also have to negotiate billing in relation to the director's or starring actors' names.

Unification into a single work. At some point in the creation of a show, it is considered to have become a single legal entity and not a collection of disparate elements such as libretto, lyrics, and music. While scenes or numbers may still be cut, added, or changed after that point, the show as a whole is considered a substantially fixed entity, and none of the writers may separate the elements again—for instance, even if the composer is unhappy with the lyricist, he may no longer withdraw the music and use it elsewhere. A collaboration agreement should determine the point at which the songs and the script become unified. For shows headed for Broadway, this point has usually been the official opening of the show in New York. But to reach Broadway has become an increasingly difficult and lengthy process. Indeed, many shows never make it. And yet, some shows do manage to obtain commercial productions elsewhere and have a life in regional or stock theaters. Therefore a collaboration agreement might specify another point at which the elements are to be considered unified—for instance, the opening of any professional production.

The length of the agreement. As mentioned above, collaboration agreements usually cover a single project. But collaborators may also wish to specify time periods within which the collaboration should reach certain goals, such as the completion of a scenario with songs, the completion of a first draft, and a full production. Any such schedule can only be a general guideline; unless you have been approached for a project by a producer with all the money in place, getting the show written and produced will inevitably take much longer than you think.

Termination. To plan for the possibility that the partnership will not work, you need to decide how to terminate the collaboration, whether the termination will be in whole or in part, and whether a new collaborator can be brought in either to replace another writer or as an addition to the writing team. Similar questions will need to be answered if a member of the team dies or becomes disabled during the term of the agreement.

We strongly suggest that you do not sign any contract or agreement unless you fully understand it and agree with it. If you don't understand it, make sure you have a lawyer or someone equally knowledgeable who can look it over and explain it to you.

Adaptation Rights

To write an original musical carries the disadvantage of having to start from scratch, so most writers tend to gravitate toward adaptations. But there are disadvantages to adaptations as well. The most important one is the rights problem: You must get permission to make an adaptation of any property that is still under copyright. Otherwise you could write a great show that no one will ever see or hear, because you did not obtain the adaptation rights to its source. This has happened many times.

A number of factors can determine whether the copyright owner will give you the adaptation rights, such as the willingness of the copyright owner to have the work adapted into a musical; the prior success of the source material and of the source's creator; the work's prior adaptation history; your individual or collective track record in creating successful shows; and the amount of revenue to be expected from the proposed adaptation.

Often copyright holders such as movie companies will refuse to give permission for certain properties, especially to unknown writers. If you do get permission, as a rule you must pay for an option that lasts for a specified period, usually between six months and two years. Depending on the source material and the demands of the copyright holders and their lawyers, the cost of the option can range from nothing to many thousands of dollars.

The first step in securing adaptation rights is to send a letter of inquiry to the copyright owners, their agents, or their attorneys, expressing your interest in adapting the work and asking about its availability. Describe your professional qualifications and the reasons why you want to do this particular project. Give a general sense of the kind of show you intend, and perhaps a sense of how the adaptation would do justice to the original work. If the copyright owner is interested, the next step is to negotiate the terms for the adaptation. You should never accept an agreement that gives the copyright owner complete control over or ownership of the project, or that defines the project as a "work for hire." (A "work for hire" is a buyout, in which you relinquish all of your rights to someone else.) To accept either clause is to give up all control over the show's content and quality, and to end up with a very bad financial deal.

Next you need to negotiate an option period. This is a mutually agreed-upon length of time in which you have the exclusive right to create the adaptation and, sometimes, to get the show produced. Often option agreements specify the maximum duration of certain phases, such as the completion of the first draft, the acceptance of the script by a producer, or the opening of a commercial production. As the adapter you want to give yourself as much time as you can—which means more time than you think you will need—but the copyright owners may try to make each period as short as possible.

You may have to negotiate the extent of the changes that can be made to the original material. As the adapter you will want to have the right to make changes as necessary, such as transposing, cutting, adding, or making other alterations in the plot, characters, and dialogue. You will also want to insert songs and dances as you see fit.

You must also either accept or negotiate the amount of the option payment. Option payments are almost always advances against future royalties to be paid to the copyright owner. As advances, options are considered *nonrefundable*—that is, the copyright owner never has to pay you back. But they are *recoupable*. In other words, when the original copyright owner's royalty from a show is calculated as a percentage of your royalties as a writer, you can often deduct the amount of your option payment from their first royalty, depending on the wording of your agreement.

Even if you can get the option and can afford it, if you do not finish the show and bring it to production within the specified time period, the copyright holder can then take back the rights, or grant them to someone else. Many writers have had projects on which they spent time and effort thus taken away from them. The cheaper and easier alternative, of course, is to choose source material from the public domain, i.e., no longer under copyright. As of this date, an artwork must usually be at least eighty-one years old to be in the public domain. Some works, like Shakespeare's plays, are obviously in the public domain, but if a work first appeared on any date after 1900, you should conduct a record search to be certain. In the United States, this is done through the Copyright Office of the Library of Congress. The search can usually be done over the Internet, but in some cases you may need either to pay the Copyright Office staff to conduct the search, or to travel to Washington, D.C., yourself.

Copyrighting Material

The word "copyright" does not simply mean "the right to copy." Exclusive rights granted to the copyright owner by the Copyright Act include the right to perform the work publicly, to reproduce the work in copies or recordings, and to distribute copies or recordings of the work to the public by sale or rental.

What this means to the creative team is that your musical will be protected under copyright law from illegal use. There are exceptions such as the "fair use" doctrine, which allows material to be quoted in reviews or for specific educational purposes, but these are limited in scope. Generally speaking, no one may use your show in any form without your permission or the permission of whoever controls the copyright, such as a publisher. Co-authors of a work are considered joint copyright owners, and any permissions must be granted by all the copyright owners, unless there is an agreement to the contrary.

Copyright protection takes effect from the moment the work is created in fixed form. "Fixed form" can mean a written or printed form, like a script or musical score, or a recording of some kind. Only the authors or those deriving their rights through the authors can rightfully claim copyright. But to be safe, we recommend that you register your work. In the United States, this is done through the Copyright Office of the Library of Congress. To register a copyright is relatively easy: Fill out the appropriate form (available by mail, or online at the Copyright Office website) and enclose a copy of the work and a check or money order for the copyright fee. While you can submit the work to the Copyright Office in the form of a recording, we strongly recommend submitting written copies of the script and score instead. In the event of a copyright dispute, the written word and note are more authoritative than any recording; they are less subject to interpretation and less dependent upon technology.

When the United States decided to adhere to the Berne Convention as of March 1, 1989, it eliminated the requirement of displaying a copyright notice. A copyright notice is still useful, however; it puts the public "on notice" that the work is protected under U.S. copyright law, it names the owner of the copyright, and it gives the first publication date.

Writing Workshops

Songwriting workshops exist at all levels, from informal gatherings of friends to workshops at a professional level. Currently the most important professional workshops for American theater writers are the Musical Theater Workshops run by the two major performance rights organizations, ASCAP and BMI. Both of these are free to writers who are accepted into the programs.

The BMI Workshop was started by the conductor Lehman Engel in the early 1960s and run by him until his death in 1982. Since then the various classes have been run by his students and their own students. It is an ongoing developmental workshop, with three different class levels: an introductory first year, an intermediate second year, and an advanced class for writers who remain for three years or longer. Classes usually meet weekly from fall through spring. At each class writers present songs, after which the teacher and the other writers give comments. (There have also been BMI librettists' workshops from time to time.) Among the shows that were created in the BMI Workshop are *Raisin, A Chorus Line, Nine, Little Shop of Horrors,* and *Once on This Island.*

The ASCAP Foundation Musical Theatre Workshop has also been running for many years. Unlike the BMI Workshop, it is project-based rather than developmental: Applicants submit, and are accepted on the basis of, specific projects rather than for long-term memberships. Thus membership changes with each workshop, which runs for several weeks or months rather than for a full season.

There are generally two kinds of workshops. In one, a teacher runs the workshop and usually guides the discussion. This is appropriate for a college or learning situation, because a knowledgeable and experienced teacher can suggest specific improvements to the writer, help other students verbalize their reactions, and make sure that criticism remains constructive. The BMI and ASCAP Workshops are both of this type. The other kind of workshop is a group of peers who meet to help each other. This can be appropriate at a certain level of expertise. In such workshops it is useful for members to take turns as moderator of the discussion. Almost any group of songwriters can set up such a workshop and help each other improve.

Workshops are invaluable for theatrical songwriters of all styles and levels of accomplishment, and any college songwriting program should include a writer's workshop, even if the students are taking private lessons. The reason is simple: They provide an instant audience. As long as at least twelve or fifteen people are in the room, they constitute an audience—and all theater writers, no matter how experienced, need an audience to tell them whether their material works or not.

It makes no difference if one person in the group dislikes you, or another is jealous of you, or yet another doesn't like anything by anyone but himself. Nor does it matter if, after the song is over, various individuals make comments that reflect a bias or agenda rather than an objective critique. Put enough people together, and when a song is performed these biases and agendas disappear: A collection of diverse individuals becomes an audience. And an audience doesn't lie. As Peter Stone said, any person in an audience may be wrong, but the audience is always right.

Auditioning a Show

Playwrights usually submit a script to producers or directors to read. This happens with musicals as well, and a "demo" (demonstration) recording is often submitted with the script. But it is also common for writers to present their show's score, or portions of their score, live.

It might seem that today, when technology allows one person to produce an entire recording or movie soundtrack in their bedroom, it is better to give potential directors or producers a fully produced recording of a new show rather than an incomplete live performance, but that is not true. So long as theater remains a live medium, the best way to present a score is live. Even if you don't sing particularly well, it is better to perform your songs than to send in a recording. Remember that you are presenting new material, not songs that people already know and love. Mediocre singer or not, you are live; a recording, no matter how lavishly produced, is dead.

Of course, if you are a truly bad singer, you would be wise to consider other options. If another member of your writing team is a good singer and performer, that is a tremendous advantage. If not, hire a few professional singers whom you know and trust. Even well-known singers are often willing to perform in this way for nominal fees; they like to be associated with new talent, and such auditions showcase their talents as well. If none of the writers can play the piano well, it is also a good idea to hire a professional show pianist to help put your songs over. Also, don't use more than the minimum number of singers needed to perform your songs. Many producers dislike the feeling of being overwhelmed by a crowd.

What should you perform in these auditions? The general consensus is to present somewhere between four and eight of the best and most varied songs from the show, unless the person for whom you are playing specifically requests more when you are done. You should not try to tell the whole story of the show, either verbally or through the selected songs; that is what the script is for. All you need is a brief description of the show to start, and a few sentences to set up each of your songs. The entire audition should last less than half an hour.

A demo recording is a necessity, but you would be wise not to make one with full production values. A fully produced recording is incongruous with a work in progress, and strikes many professionals as overkill. It is much more effective simply to have good singers and good piano accompaniment that are well recorded, without excessive reverberation or any pop-music effects.

If your show makes it to the point of a commercial production, you may find yourself doing another type of audition, the backer's audition. These are usually presented by producers for potential investors. Like other auditions, backer's auditions should never present the show in its entirety. They should consist mostly of songs and should last no more than forty-five minutes. Sometimes the director or star, if already committed to the project, will also perform or help "sell" it at the backer's audition.

Agents

Writers' agents ordinarily get up to fifteen percent of any money their clients earn as writers. A common misconception holds that agents go out and find work for their clients. This rarely happens for writers; they must find their own productions.

Why pay an agent fifteen percent, then? Agents make the deal. They negotiate it for you, they nail down the contract, and they look out for your interests. In addition, many producers will not look at new shows unless they are submitted through an agent.

How do you get an agent? Unlike the film industry, where you need a track record in order to get an agent and you need an agent in order to get a track record, the theater is a little more open. Since many producers and theaters are willing to consider shows submitted without an agent, it is possible to develop a track record and then approach agents based on that track record. And some agents are willing to consider representing unproduced writers, knowing that a relatively small investment of time might possibly reap large rewards in the future. While these names change every year, current lists—like those of producers and regional theaters—can be found through the Dramatists Guild or various publications, most of which are available at stores such as the Drama Book Shop (in New York City or online). The *Writer's Market* and *Song Writer's Market* books, updated annually, also contain some of this information.

Production Possibilities

READINGS AND STAGED READINGS

A theater work truly comes to life only when it is performed on a stage with full production values. But beginning writers cannot always get such performances, especially in the early stages of a new musical. While they are not actual productions, both readings and "staged readings" can still be helpful in giving writers a chance to hear and assess what they have written, especially if professional actors do the reading. Most of Stephen Sondheim's shows, for example, were tested in readings before going into a full production, and he found the readings extremely valuable.

The difference between a reading and a staged reading is that in a reading the actors usually sit and read from their scripts. In a staged reading, the actors walk around and follow the stage directions to some degree. Sometimes they perform all or part of their roles from memory. In readings, the songs are often performed by the writer or writers at the piano, while in staged readings they are usually sung by the actors. Also, readings are usually done in private for the benefit of the writers or the production team, while staged readings are usually done for an audience.

WORKSHOP PRODUCTIONS

Developmental workshop productions are sanctioned by Actors' Equity Association, the actors' union, for developing new works. In Equity workshops actors make lower salaries than usual. In exchange, they split a percentage of the gross box office and subsidiary rights, and also get transportation expenses and health and pension benefits. Workshop performances are only for invited guests.

Shows like *A Chorus Line, Dreamgirls*, and *Sunday in the Park with George* were originally developed and presented in workshops before going on to Broadway, and there are now theater companies in many cities that specialize in workshop productions. In recent years, however, workshops have tended to deviate from their original purpose. Most often they are glorified backer's auditions, rather than places where the creative team works to improve the show. Many shows have emerged from workshop productions no better than they were before, and as a result, many that raised the money for production failed anyway.

SHOWCASES

Actors' Equity Association developed its Showcase Contract to help foster inexpensive shows for limited runs, usually two weeks. According to the Equity contract, showcases are meant for venues with 99 or fewer seats. In showcases, unlike workshops, actors usually work without pay. Unlike a workshop production, a showcase is open to the public and can sell tickets. Showcases are also often used to find a producer for a full production. If a producer is already interested, then like workshops, showcases are often used as backer's auditions, to raise money for an open-ended commercial production.

EDUCATIONAL PRODUCTIONS

It is often difficult for writers, especially new ones, to get any sort of professional production. In this case they might consider having their show produced by a college or high school. Performances at the college and university levels can vary from atrocious to near-professional quality. Sometimes an ambitious college program will even pay for well-known actors or directors to work on a new show. High school groups, on the other hand, can make up for their lack of training and experience with energy and enthusiasm.

The educational market for older shows is large, but most schools rarely produce new shows by unknown writers unless the writer has a personal connection with the school. Some university drama and musical theater programs provide opportunities for the production of new works through competitions, workshops, and theater festivals.

An educational production can point up flaws in a show, allowing writers to fix them before the show goes further. On the other hand, budgets are often severely limited, and the subject matter is often limited to what students can understand and perform. Also, no matter how much enthusiasm they exhibit, nonprofessionals will give a nonprofessional performance, and if the performances are very bad, they will not help you find your own mistakes.

REGIONAL PRODUCTIONS

Every major city and many smaller communities in the United States have developed their own regional theaters, some of which perform new musicals. Many of these theaters currently belong to the National Alliance for Musical Theatre, an organization that allows them to share the costs of producing new shows.

Dinner theater is a special category of regional theater. Most dinner theaters do not produce new shows; a few occasionally do.

FULL-SCALE COMMERCIAL PRODUCTIONS

If you have reached the point of a full commercial production, you have an entirely different set of concerns from those you had before, and they lie beyond the scope of this Appendix. The most important things for you are a good agent, a good lawyer, and a Dramatists Guild contract.

APPENDIX D

REQUIRED READING
(AND LISTENING)

Anyone interested in writing musical theater should be familiar with the following shows and songwriters. To try to create a work in a particular artistic genre without studying the best works in the field is like starting a trek through the desert without bringing water: There is a theoretical chance that you might make it, but why start off with all the odds against you?

Classic Shows of the Golden Age

The following musicals, listed in chronological order, are the classics. They constitute a canon of high quality in both libretto and score. They are all worth studying in detail. Remastered versions of the original cast recordings of all these shows are usually available from music stores and libraries.

Rodgers and Hammerstein. *Oklahoma!*
Rodgers and Hammerstein. *Carousel.*
Fields, Fields, and Berlin. *Annie Get Your Gun.*
Lerner and Loewe. *Brigadoon.*
Rodgers, Hammerstein, and Logan. *South Pacific.*
Spewack, Spewack, and Porter. *Kiss Me, Kate.*
Swerling, Burrows, and Loesser. *Guys and Dolls.*
Rodgers and Hammerstein. *The King and I.*
Lerner and Loewe. *My Fair Lady.*
Loesser, Frank. *The Most Happy Fella.*
Laurents, Sondheim, and Bernstein. *West Side Story.*
Laurents, Sondheim, and Styne. *Gypsy.*
Burrows, Loesser, et al. *How to Succeed in Business without Really Trying.*

Shevelove, Gelbart, and Sondheim. *A Funny Thing Happened on the Way to the Forum.*
Stein, Harnick, and Bock. *Fiddler on the Roof.*

Other Important Shows

Most of the following shows appeared either before or after the "golden age." If they are not quite the equals of the shows listed above, they come very close.

Hammerstein and Kern. *Show Boat.*
Kaufman, Ryskind, and the Gershwins. *Of Thee I Sing.*
Harburg, Saidy, and Lane. *Finian's Rainbow.*
Willson, Meredith. *The Music Man.*
Jones and Schmidt. *The Fantasticks.*
Masteroff, Harnick, and Bock. *She Loves Me.*
Masteroff, Kander, and Ebb. *Cabaret.*
Furth and Sondheim. *Company.*
Goldman and Sondheim. *Follies.*
Wheeler and Sondheim. *A Little Night Music.*
Maltby and Shire. *Starting Here, Starting Now.*
Wheeler and Sondheim. *Sweeney Todd.*
Kleban, Hamlisch, et al. *A Chorus Line.*
Ashman and Menken. *Little Shop of Horrors.*
Maltby and Shire. *Closer than Ever.*
Klein, Price, and Kleban. *A Class Act.*

Important Songwriters

The following is an alphabetical list of composers and lyricists who produced theater songs of high quality in the first half of the twentieth century, songs that have survived the shows in which they first appeared. Collections of theater songs by any of the following writers are worth studying. The same is true of collections by any of the other songwriters mentioned above.

Harold Arlen
Irving Berlin
B. G. "Buddy" De Sylva, Lew Brown, and Ray Henderson
Vernon Duke
George and Ira Gershwin
E. Y. "Yip" Harburg
Jerome Kern
Johnny Mercer
Cole Porter
Richard Rodgers and Lorenz Hart
Arthur Schwartz and Howard Dietz
Kurt Weill
Vincent Youmans

INDEX